HOT TOPIC

Global warming and the future of New Zealand

GARETH RENOWDEN

DEDICATION

For Tim and Emma.
It's your future.

www.hot-topic.co.nz

Published by AUT Media
PO Box 7125, Wellesley St
AUCKLAND
Email: info@hbmedia.co.nz
www.autmedia.co.nz

Copyright © Gareth Renowden 2007
First published 2007

The author has asserted his moral rights in the work.

This book is copyright. Except for the purposes of fair reviewing, no part of this publication (whether it be in any eBook, digital, electronic or traditionally printed format or otherwise) may be reproduced or transmitted in any form or by any means, electronic, digital or mechanical, including CD, DVD, eBook, PDF format, photocopying, recording, or any information storage and retrieval system, including by any means via the internet or World Wide Web, or by any means yet undiscovered, without permission in writing from the publisher. Infringers of copyright render themselves liable to prosecution.

ISBN 978-0-9582829-0-1

Designed by HB Media
Edited by Bradstock & Associates, Christchurch

CONTENTS

Foreword by Dr James Renwick	5
Preface	7
Introduction	9
1: Living in a greenhouse	15
2: The climate system	27
3: The state of the science	39
4: The outlook for New Zealand	53
5: Impacts: the good and the bad	63
6: Sinking or burning: our Pacific neighbours	77
7: Warming in the wider world	85
8: Facing up to the inevitable	91
9: Cooling the future	115
10: A low-carbon New Zealand	135
11: The big picture	153
12: The way forward	163
Appendix: The sceptical view	169
Appendix: Notes and resources	181
List of acronyms	191
Glossary	192
Acknowledgements	195
Index	197

FOREWORD

It's my pleasure to provide a preface to this important book by Gareth Renowden. Climate change is a crucial issue facing humanity in the 21st century, one that challenges the very foundations of modern society, yet one that has not been addressed in this way for the New Zealand reader before the publication of *Hot Topic*. Outlining the key science findings on climate change, as summarised by the recent Assessment Reports of the Intergovernmental Panel on Climate Change (IPCC) and illustrated with personal experiences and local knowledge, *Hot Topic* makes the science, the impacts, and the necessary action accessible and intelligible to all.

Possible futures under climate change range from the relatively benign to the positively scary. New Zealand's location in a very large and slow-to-warm ocean means the rate of change here may be less than in other countries. So, some sectors of our primary industries might fare well over the next few decades and we may be able to capitalise on longer growing seasons and lower frost risk, for example. Yet even here, changes in water availability (a key theme for much of the globe in coming decades) will place added stresses on existing land uses. And what will more severe impacts do to the economies of our trading partners, to world trade and to our export sector? If change is relatively benign here, but is not elsewhere, there may be a lot of people who want to move to New Zealand, whether we like it or not.

New Zealand is likely to face major political, economic and biosecurity issues as patterns of world trade change and new arrivals (including tropical pests and diseases) queue up to come ashore. Conditions in neighbouring Pacific Island states might be a special cause for concern for us, as clearly described in this book. How we cope with such challenges will shape our future social and political landscape for many decades to come.

The threat of climate change is as insidious as it is alarming. Our short memories render us largely unaware of the gradual but cumulative changes wrought by steadily increasing greenhouse gas levels. Apparently small-sounding changes in average conditions often go with large changes in the risk of extreme events. We need to take stock of the big picture to understand the real situation and to appreciate the urgent need for action. *Hot Topic* gives us that big picture, and brings it home to New Zealanders.

And why should we take action? New Zealand accounts for a tiny fraction of global greenhouse gas emissions, so what can we as a nation of only 4 million do? The answer is that the whole globe is made up of many, many groups of 4 million people, be they nations or merely suburbs. Solutions must begin at home – all around the world. New Zealand is known globally as the "clean green" country. If we can't take positive action, who will? It is our responsibility as stewards of our part of the planet to do what we can to preserve the Earth system as it is, for the sake of our children's future. If we care about the survival of native flora and fauna, the safety of coastal communities, the health of our economy and the quality of our lives, then we have an obligation to take the problem very seriously.

Hot Topic lays out the facts and details options for the future. Gareth pulls no punches yet remains optimistic and spells out what we can do as a nation, and as individual citizens of the world, to make a positive difference. Read this book.

Dr James Renwick
Climate Research, NIWA
IPCC AR4 Lead Author

PREFACE

I've always been interested in weather. During my childhood, I spent a few years on the small Scottish island of Tiree. I remember my father taking meteorological observations as part of his job as airport manager. I remember the planes and the maps and charts, but more than anything I remember the weather and the beaches. Tiree is warm and windy – it is one of the windiest places in the British Isles, so windswept and flat that no trees grow there.

People who are interested in weather tend to be interested in extremes. I remember the gales blowing over the island, the surf smashing on the empty beaches, the sand drifting over the roads, and standing on the low stone wall outside our house with my school blazer spread open, waiting for a blast of wind to blow me into a pile of straw. While I was there, the strongest gust was around 110 mph – nearly 180 kph. It blew the doors off the hangar and bent the anemometer, so it may have been even stronger.

Big weather events capture the imagination – heavy snowstorms, damaging floods, gales that blow down forests. Climate, which is what weather averages out to over time, seems intrinsically less interesting. It's more about statistics than excitement. Climate would be a dull subject if it weren't for the fact that when it changes, it affects us all.

I've followed the development of climate science with interest since my student days in the early 1970s. My own background is life sciences rather than atmospheric physics or paleoclimatology, but I can remember when there was concern that we might be heading towards another ice age (which, being a keen skier, I thought might be quite interesting), and when it became clear that global warming was likely to be the bigger problem. In recent years, I have been fascinated by the way that developments in computers and communications have combined with space technology to give us a new and highly detailed view of the way our planet works. I can visit the NASA *Earth Observatory* website and look down on fires in Australia and dust storms over China, or click on *Cryosphere Today* for a look at how much sea-ice surrounds Antarctica. From my office, I command a view of our planet that scientists a few decades ago could only dream about. It may not have the immediacy (or excitement) of circling the planet in the International Space Station, but because I am looking through electronic sensors as well as my eyes, I can take in much more than any astronaut. My senses are working overtime, giving me a picture of a planet where human impact is becoming only too clear.

My wife Camille and I bought our little farm in the Waipara Valley in the mid-1990s, and we try to grow olives, grapes and truffles. The local climate and soil – what the French call the *terroir* – is (I hope) perfectly suited to producing world-class pinot noir, high-quality olive oil, and highly valuable, aromatic black truffles. Every year, our plants respond to the weather by changing the timing of various milestones – flowering, fruiting, achieving optimum ripeness. The plants integrate that weather, pull together all the rain, heat and wind that modulate their growth, and express the season in their fruit. If our climate changes, so the taste of that fruit will change. The local climate determines the quality of my crop, and so I take the potential for change rather seriously. This book is my attempt to find out what those changes might be, and what we can do as individuals and as a nation to respond to the challenge climate change poses.

> **A note on terminology:** Throughout this book I use the terms "global warming" and "climate change" more or less interchangeably. Strictly speaking, I am discussing the changes in climate that result from the warming effect of increased greenhouse-gas levels in the global atmosphere, but that is rather a mouthful to repeat on a regular basis. Both terms have, however, acquired a certain amount of political baggage. "Global warming", because it is a statement about what's happening, admitting no other possibility, has been seen to be a more contentious phrase than "climate change". After all, climate changes can go down as well as up, like shares and interest rates. I think, and I hope this book will demonstrate, that the reality of global warming is beyond dispute, and so I will use the phrase quite happily alongside "climate change". But you will know what I mean.

INTRODUCTION

Our climate is changing. New Zealand is getting warmer. The whole world is warming, and it will continue to heat up for decades to come. We know the cause, and the choices we make over the next few years will determine whether we (and the world) can cope gracefully, or whether we run into severe trouble. *Hot Topic* sets out to explain what we currently understand about the science of this global warming, and to explore what the resulting climate change means for New Zealand. We will have to adapt to the warming that's inevitable, and we will have to participate in global efforts to reduce the potential for damaging change. That has significant implications for the way we live and work, and our relationship with the rest of the world. The way we respond to this challenge is going to be a big part of the rest of our lives.

Global warming and the climate change that it causes have become controversial subjects. Over the past 20 years, a huge scientific effort has succeeded in fingering the prime suspect – CO_2 emissions, mainly from burning fossil fuel – and has suggested that the warming world won't be a pleasant place in which to live. This news has not been welcomed by companies that make their profits by supplying or burning fossil fuels. Large corporate interests – especially in the USA – have been keen to downplay the issue, and in some cases have funded campaigns to deny that the problem exists. These campaigns have found political support beyond the USA. Politics and science have become intertwined in ways that are difficult to untangle, and which make action on climate change harder to achieve. In this book, I want to try to take the politics out of the science and put it where it belongs – in deciding what our response to the problem should be.

The main tactic used by those who want to minimise the perception of climate change as a pressing problem is to insist that there is still a real debate about the causes and extent of global warming. As far as they're concerned, if we are still debating the basics of the problem – whether it exists, or how big an issue it's likely to be – then we don't need to do anything about it. Business can continue as usual. Many people find that a comforting message – but, unfortunately, it's framing the issue in a very misleading way. There is no longer any real debate about the cause of global warming, as we shall see in the first part of this book. It's us. There is a lot of debate about just how bad it might be, how the climate system will respond and what will happen to our way of life, but the central message is no longer controversial. It's established fact: the amount of CO_2 and other greenhouse-gases in the atmosphere is on the increase, owing to human actions, and it is warming up the planet.

It's also clear that some climate changes will not be subtle. The Arctic Ocean could lose most of its summer ice in my lifetime (I'm 52 as I write, and not overly optimistic about my lifespan). It's worth noting that the difference between a summer we think of as hot and one that's just normal can be as little as 1°C in the *average* temperature. It's a little number, but it means a lot. So when we talk about global temperatures increasing by a couple of degrees, we're actually talking about a big climate change. Hidden in that global average will be places that warm up a lot more than others. For example the poles – especially the Arctic – will warm much more than the average. New Zealand is relatively lucky. Stuck in the middle of the large Southern Ocean, which will take a long time to warm up, we should warm more gently than most. Adapting to a changing climate will be easier here than in many other places. One scientist told me that adapting to a warmer New Zealand is likely to require fewer land-use changes by Kiwi farmers than the Rogernomics reforms of the 1980s.

Climate change is an external threat to our way of life. It's like being confronted by an armed and aggressive enemy: ignoring the problem will only make it worse. The rational response is to do as much as we can to learn how well-armed our enemy is, what he is going to do, and then to defend ourselves as best we can. In terms of global warming, climate science is our intelligence-gathering. It tells us what the problem is, and gives an idea of how bad it might be. It also points us at possible solutions. Ignoring this information isn't good policy. Shooting the messenger certainly isn't going to make the bad news go away. We have to focus on our strategy and tactics, decide how we're going to respond and how we're going to defend ourselves.

How would all the countries of the world react if we discovered an asteroid on a collision course with Earth – an asteroid big enough to wipe us all out in the same way that the dinosaurs died? Would we set aside our differences and work together to try to find a means to prevent the collision? I think we might agree that it was worth a damn good try. Climate change is not as obvious, dramatic or final as an impending asteroid impact, and only a very few people think that it will bring the human race to the edge of extinction, but the principle is the same. It's a global problem that requires a global solution.

Getting that global cooperation, however, is not a trivial undertaking. No one country can solve the problem by itself. New Zealand's greenhouse-gas emissions are less than 0.5 percent of the global total. Even if we were to cut our emissions by 90 percent it would have no measurable impact on the planet's atmosphere. In global-warming terms, as in so many other things, we have to follow someone else's lead. It is, however, very much in our interests that the rest of the world does something to sort the problem out. We will be bearing at least some of the costs of change in our own country, and we are extremely vulnerable to the impacts elsewhere. If any of our major trading partners gets hit badly, we'll feel that in our pockets. By adopting the same multilateral approach we take in world trade negotiations, we can at least contribute to meaningful progress. The international progress we've already achieved, from the United Nations Framework Convention on Climate Change to the Kyoto Protocol, may not be universally popular, but is an important first step down that road.

If New Zealand does nothing to cooperate with the rest of the international community on climate change, we will be at great risk of other countries refusing to trade with us, or erecting trade barriers against our products. In late 2006, French prime minister Dominique de Villepin floated a proposal that Europe should charge a carbon tax on products from countries that were not participating in international agreements to reduce greenhouse-gas emissions. That proposal didn't gain much support this time around, but it would be a brave government that ignored the obvious message: there will be sticks as well as carrots in a low-carbon world.

Beyond being a good global citizen, New Zealand has another role to play in finding solutions. Our emissions "profile" – the mixture of greenhouse-gases we emit each year – is unusual in world terms, because around half of our emissions come from agriculture. Methane and nitrous oxide, both powerful greenhouse-gases, are produced by pastoral agriculture, in the stomachs of sheep and cows, and through fertiliser use on the grass they eat. In most countries, power generation, industry and transport are far more important emitters of greenhouse-gases, particularly CO_2, so that's where they focus their attempts to reduce emissions. It's in New Zealand's interest to work hard at the agricultural problem, because if we can find ways of cutting those emissions it will not only help us to achieve any emissions-reduction targets we set ourselves, but will give us expertise that can be used and sold around the world. We are known worldwide as leaders in agricultural development and innovation, and we can become leaders in solving the environmental problems associated with large-scale farming.

Cutting our emissions is going to be important, but adapting to climate change as it happens will be New Zealand's biggest challenge. We need to find out as much as we can about what changes to global climate mean on a local scale. If the world warms 2°C on average, what does that mean for summers in Hokitika or Whangamata? Will rainfall increase or droughts become more common? Will there be more floods, or more damaging storms? Will rising sea levels flood our beachside baches? Would you spend hundreds of thousands of dollars on a sea-front home if you thought it might be vulnerable to storm surges in a few decades' time? Defining these risks and factoring them into our planning processes are not glamorous, but are essential.

It's more than a matter of adapting to the changing physical environment around us. There is also the question of how New Zealand society will adapt to a low-carbon world. Prime Minister Helen Clark has stated that New Zealand should have "carbon neutrality" as a long-term goal. This means putting only as much carbon into the atmosphere as we take out of it each year by growing trees and other plants. That's ambitious, but not impossible. New Zealand is probably as well placed as any country in the world to be carbon-neutral, mainly because of the substantial hydro and thermal power-generation system we already have, and because we're very good at growing trees. But it will require a lot of changes: more electricity generation from renewables such as wind, hydro, thermal, tidal and solar sources (perhaps with a shift away from large generating plants to small-scale micro-generation); moving away from using oil as a transport fuel and making more use of biofuels and biodiesel (with all that implies for our agricultural crop mix); more use of public transport in our cities; promoting the use of low-emission and electric vehicles and using solar power to warm our houses and heat water.

Beyond the purely local issues, future governments will also need to devise strategies for dealing with new realities. If rising sea levels pose a significant problem for our Pacific Island neighbours, how will we handle an influx of climate-change refugees? Will our armed forces have to patrol our borders in the same way that Australia patrols the waters to its north? If the world moves towards eating local food – reducing the "food miles" that have already emerged as an issue for our agricultural products in Europe – then we will need to have a response. If international air transport becomes more expensive because of its impact on climate, how will we cope with a decrease in tourism and general business travel?

These sorts of changes will require government action and there will be plenty of arguing about the best way to achieve the bigger and more difficult goals. This requires New Zealand's leaders to take the problem seriously. However we still have politicians who are openly sceptical about global warming, and there is no broad agreement among the political parties as to where we should be heading, nor are there any targets to meet. A sense of urgency is lacking.

This was brought home to me forcefully during a recent trip to Europe. As I arrived in London, the walkway from the plane to the immigration hall was lined with advertisements declaring BP's commitment – as one of the biggest oil companies in the world – to dealing with climate change and moving to a low-carbon future. Later, driving through Spain, I was amazed by the number of giant wind turbines rotating (rather majestically, I thought) on hills as though waiting for a high-tec Quixote to tilt at them. And in France, on an unscheduled detour around a nuclear power station (France gets 80 percent of its electricity from nuclear generation), I was even more amazed when a New Zealand colleague quite emphatically denied that global warming was a problem, or that we should do anything about it. The contrast was stark. While New Zealanders were still coming to terms with reality, the people of Europe were focusing on solutions to a problem they knew was real. New Zealand has some catching up to do.

1

Living in a greenhouse

The planet Earth is a ball of rock with a thin skin of water and air orbiting the sun at a distance of 150 million kilometres. The surface of the sun is hot, roughly 5500°C, and it radiates a huge amount of energy out into space. A small amount of that energy hits the side of the Earth facing the sun, and heats it up. To that energy, we owe our existence. The heat keeps the planet warm enough to have liquid oceans. Plants take the sunlight and use it to power the production of sugars. Animals eat plants, and other animals. Fungi and bacteria recycle dead things. A vastly complicated web of life, from single-celled organisms to complex multi-cellular beings like us, lives in the planet's thin skin – the biosphere.

If the Earth had no atmosphere, the average surface temperature would be a distinctly chilly minus 18°C. However, the envelope of gas that makes up the atmosphere captures some of the heat and warms the planet up to a global average of about 14°C. In simple terms, the solar energy arrives mostly as light – short-wavelength radiation. This heats up the Earth's surface, which then radiates longer-wavelength thermal radiation back into space. Some of the gases in the atmosphere are more transparent to light than they are to the heat, so they capture some of that energy and direct it back to the ground. This is called the "greenhouse effect"* and the gases in the atmosphere that keep us warm are called greenhouse-gases. The most important of these are water vapour, CO_2, methane, ozone, nitrous oxide and a group of substances called halocarbons (which also cause the ozone hole). Water vapour has the biggest effect, followed by CO_2, methane and nitrous oxide, although molecule-for-molecule, methane and nitrous oxide pack a much bigger punch than CO_2 and water vapour.

At the distance the Earth lies from the sun, the amount of solar energy averages 1366 watts per square metre (Wm^{-2}), as measured by satellites. Allowing for the fact the Earth is a sphere and not a flat disc, that translates to about 342 Wm^{-2}. Again, that figure is an average, and what actually arrives at the surface depends on latitude, time of day, season,

---

* This term is in fact misleading, because greenhouses don't actually work the same way. They heat up because the sun heats the surfaces inside, which then warm the surrounding air. But this warmed air can't blow away or escape – it's trapped by the glass.

atmospheric conditions and weather. Averaging across seasons and latitudes and reflectivity (called albedo), roughly 235 Wm^{-2} is absorbed by the Earth. This is called the insolation. This heat input is enough to raise the temperature of the Earth's surface to minus 18°C, but the effect of all the greenhouse-gases working together add another 324 Wm^{-2}. The extra heat boosts the temperature by 32°C, enabling the planet to have liquid oceans, and creating the conditions within which all life has evolved (see Fig 1, p. 97).

To maintain a steady temperature, the amount of energy arriving at the planet has to match the energy lost as radiation back out into space. If it radiates more than it receives, it cools down. At a local level, this is what happens as day turns to night and when seasons change. Water can absorb a lot of heat and only releases it slowly, so the Earth's oceans, which cover 70 percent of the planet, act as a heat sink and store a huge amount of energy. They absorb heat during the warm seasons and give it up slowly in the colder seasons. That's why the hottest summer months occur after the longest day, and why the coldest winter months occur after the shortest day. The sea starts to cool only when the sun's power wanes in autumn, and carries on cooling over winter until the strengthening spring sun can add heat faster than it is being lost. The oceans have a buffering effect on the land they surround, which is very important for an island nation like New Zealand, helping to protect us from extremes of heat and cold.

The latest measurements suggest that the Earth is currently absorbing about 0.85 Wm^{-2} more than it is radiating back into space. That's why we're warming up. It may not seem like a big number, but when you multiply it by the huge surface area of the planet it amounts to 30 times as much energy as the whole human race produces and consumes each year. And when we look at the amounts of greenhouse-gases in the atmosphere, we find that since the middle of the 19th century the level of CO_2 has increased by 36 percent, and methane by 150 percent.

There is a lot of evidence that the world is getting warmer. Eleven of the last 12 years (1995–2006) are among the 12 warmest ever recorded. Systematic and accurate recording of temperatures in New Zealand began in the middle of the 19th century. Since 1900, the average temperature has risen by about 0.9°C, but the period around the beginning of the last century was the coldest in the New Zealand record, about 0.3°C cooler than the 1870s (see Fig 2, p. 97).

There are two principal collections of global temperature data, one at the Goddard Institute for Space Studies (GISS), a division of NASA, the other at the Hadley Centre in the UK. They differ slightly in the way that they put the figures together, but both show the same thing. Cool in the late 19th century, followed by warming until the 1940s, then a period of slight cooling until the mid-1970s and a steady warming ever since. According to the GISS dataset, the world has warmed by almost 0.6°C since the 1970s. The year 2005 was the warmest ever recorded, just beating 1998, a year when a very strong El Niño effect in the Pacific caused global temperatures to spike. The Hadley Centre puts 2005 very slightly below 1998 – the second hottest year in the record. The 10 warmest years in the Hadley Centre records have all occurred in the last 12 years. The trend is strongly upwards at around 0.2°C every decade (see Fig 3, p. 98).

To appreciate how significant this is, we need to put the present warming into an historical perspective. The longest-running set of temperature data in the world comes from Britain. The Central England Temperature (CET) record begins in 1659 as a series of monthly average temperatures. From 1772 onwards, the CET consists of daily mean temperatures. The CET figures are not a global average but echo the global picture, showing that 2006 was the warmest year in England since the series began, and that six of the warmest 10 years have been in the last decade.

To look back before the 17th century, we have to find things that can stand in for thermometers – temperature proxies, as they're known. Over the last 50 years, ingenious scientists have discovered lots of them, from the width of tree rings to the ratio of oxygen isotopes in cores drilled in the Antarctic or Greenland ice sheets. Studying the ancient climate (paleoclimatology) is a very important part of developing an understanding of how the global climate system works. If we can

> **El Niño/Southern Oscillation**
>
> The El Niño/Southern Oscillation (ENSO) is a change in the patterns of ocean temperature and atmospheric pressure in the tropical Pacific which has a marked effect on weather around the world. Normally the easterly trade winds push warm water from the seas off America to the west. This allows nutrient-rich cool water from the deep ocean to rise up to the surface off the coast of central and southern America. This nutrient-rich water is essential for the growth of plankton, the base of the marine food chains. In an El Niño year, the trade winds weaken and warm water stays in the eastern Pacific. There is less upwelling of cold water and the fisheries of Ecuador and Peru suffer as a result. El Niño means "the boy child" in Spanish and was so named because it often begins around Christmas.
>
> The opposite of El Niño is La Niña – "the little girl". During the La Niña phase of the ENSO, easterly winds are stronger than normal, building up a larger-than-usual pool of warm water in the western Pacific, while ocean temperatures in the east are lower than normal. These sea temperature changes drive changes in weather patterns by affecting the positions of high and low pressure systems.
>
> ENSO's effects on weather in New Zealand depend on the time of year. A summer El Niño can bring more westerly winds, leading to drought on the east coast, and it can be cooler than normal. La Niña usually brings more northeasterly winds, and increased rainfall in the northeast of the country.

discover how the world's climate has changed and figure out what triggered those changes, we can learn how sensitive it is to the factors that drive change, as well as how it might behave in the future.

Studying long cores of ice from Antarctica and Greenland provides compelling evidence that we lived in an ice age. By measuring the ratios of oxygen isotopes in the ice we can tell the temperature at the time the snow fell. By analysing the air trapped in bubbles in the ice layers, we can determine the composition of atmosphere at the time. Antarctic cores now provide records that go back more than 800,000 years. Over that time, the world has dipped in and out of ice-age conditions roughly every 100,000 years, with much shorter, warmer periods in between, called interglacials. During each ice age, huge ice sheets formed over North America and Europe, and New Zealand's glaciers were much larger, extending out to what are now the Canterbury plains and the Tasman Sea.

The records of temperature and CO_2 show that ice ages tend to end comparatively suddenly, leading to a relatively short period of mild temperatures (the interglacial) followed by a slow descent back into a full ice age. Around 13,000 years ago the last set of ice sheets began to retreat as the climate warmed. Humans moved from living in hunter/gatherer groups to more organised societies with agriculture, cities and civilisations, and made their mark on the world. This climatic period is called the Holocene, and it has been kind to our species.

The ice-core data tell another interesting story. Very precise measurements of CO_2 in the bubbles trapped in the ice show that throughout the long cycles between ice age and warm period, the CO_2 levels move in step with changes of temperature. During interglacials, CO_2 reaches about 280 parts per million (ppm), but it drops to under 200 ppm in the depths of an ice age. The ice cores also show that the CO_2 level in the atmosphere in the middle of the 19th century was about 280 ppm – much lower than today, and commonly referred to as the pre-industrial level of CO_2 (see Fig 4, p. 98).

The trigger that turns ice ages on and off is thought to be periodic changes in the Earth's orbit around the sun. These changes, called Milankovitch Cycles (after the Serbian scientist who first worked out their intricacies), slightly change the amount of sunlight reaching different parts of the Earth at different times of the year. At times when the northern hemisphere receives less sunlight in summer, ice sheets start to form over the continents around the North Pole. There's an interesting lesson in the ice age/CO_2/Milankovitch relationship. Orbital variations have only a tiny effect on the average amount of solar energy reaching the Earth, but they can tip the climate from ice age to warm interglacial and back again. The climate system seems to respond to small changes very quickly. The changes we're making to the heat budget of the planet by increasing the amounts of greenhouse-gas in the atmosphere are much larger than the Milankovitch effects, and the rate of change in temperature is already very fast.

Since the start of the Holocene, the global climate has been warm, but not uniformly so. There have been times when some regions have been warm and others cold. Glaciers in the Southern Alps have advanced and receded. Over the last 1000 years, Europe has experienced the Medieval Warm Period (800–1300AD) followed by a Little Ice Age, which ended in the 19th century. Studies of the climate of the last couple of thousand years using a range of temperature proxies, including tree rings, suggest that modern temperatures are warmer than at any time in the last 1300 years – possibly since the beginning of the Holocene – and they could be within one degree of the highest temperatures in the last five million years.

The temperature record is not the only evidence of widespread warming. Over the last 100 years New Zealand has lost about a third of its area of glaciers, and in the last 50 years, a quarter of the total mass of ice. Sea ice at the North Pole is not only shrinking in area, but getting thinner as well. The summer ice minimum is declining by about 9 percent every decade, and by 2040 it may be possible to sail all the way to the North Pole in summer. Ice shelves are breaking up in Antarctica and Northern Canada, and satellite measurements show that the amount of ice in the Antarctic and Greenland ice sheets is decreasing. Greenland's ice sheet could be losing as much as 330 km^3 of ice every year, and this rate has doubled over the last five years.

The Intergovernmental Panel on Climate Change's (IPCC) latest report describes the evidence for warming as "unequivocal". The evidence of warming can be clearly seen in the global temperature record, melting ice, rising sea-level and warming oceans. Average Arctic temperatures have increased twice as fast as the global average over the last 100 years. Permafrost is melting – the top layer has warmed by as much as 3°C since the 1980s. The frequency of heavy rain has increased over most land areas. Temperature extremes have shifted: there are far fewer cold events and heat waves are more frequent. We are living in a warming world.

The role of greenhouse-gases in keeping the planet warm has been understood for more than a century. In 1827 the French mathematician and scientist Joseph Fourier published an essay suggesting that the Earth's atmosphere might trap heat from the sun. The Irish natural philosopher John Tyndall built on that idea, and in 1859 showed that water vapour and CO_2 could block thermal radiation. Tyndall was also interested in glaciers (there are glaciers named after him all over the world, incluing New Zealand) and intrigued by evidence that in the past Europe and North America had been covered by huge sheets of ice. Scientists at the time believed that climate was not prone to great or sudden change, but if glaciers had once extended down from the Alps into the countryside (and even covered great tracts of Britain) then the climate must have been much colder. Working out how that might happen was one of the most intriguing puzzles in the Earth sciences, and still motivates research today.

The man who first tried to work out the relationship between the amount of greenhouse-gas in the atmosphere and the temperature of the Earth was a Swede, Svante Arrhenius. He wanted to see if reducing the amount of CO_2 in the atmosphere could bring about an ice age, so he constructed a mathematical model – a system of equations and relationships – describing how water vapour and CO_2 in the atmosphere would affect the temperature at the surface of the Earth, using early measurements of how the gases transmitted radiation. He then laboriously calculated the effect across the globe – a huge undertaking in the days before calculators, let alone supercomputers. In 1896 he published his findings: if the level of CO_2 in the atmosphere halved, the temperature in Europe would fall by 4–5°C. Then, after a colleague asked what would happen if CO_2 levels increased, he worked out that the global temperature would increase by 5–6°C if CO_2 doubled. Modern calculations suggest that doubling CO_2 will increase global temperatures by 2–4.5°C, so he wasn't too far off.

Unfortunately for Arrhenius, his work didn't catch on. Most scientists believed other research, later shown to be highly flawed, which suggested that increasing the amounts of both gases wouldn't make much difference to the amount of heat they trapped. However, a few people kept the idea alive, notably a British steam engineer called Guy Callendar. In the late 1930s he became intrigued by suggestions that the climate had warmed over recent years, and compiled a series of temperature records that confirmed this. He then looked at the CO_2 level in the atmosphere over the same period, and found it had increased by 10 percent. He put these two things together, developed a rough model of the greenhouse effect, and calculated that the increase in CO_2 could account for the temperature rise. However, his work didn't win him many fans, because most meteorologists and physicists still believed the flawed data on how greenhouse-gases behaved, and thought the measurements of CO_2 levels were inaccurate (they were). They also pointed out that CO_2 dissolves in seawater, so the oceans might soak up all the CO_2 being added to the atmosphere.

In the 1950s, advances in technology and computing made it a lot easier to do the complex calculations involved in describing how the atmosphere works. The Cold War drove a lot of technological development, including significant advances in atmospheric physics and better measurements of radiation absorption by the gases in the atmosphere. It became clear that increasing amounts of CO_2 and water vapour would trap more heat at the Earth's surface. At the same time, it was discovered that CO_2 did not dissolve in the oceans as easily as had been thought. There was renewed interest in the amount of CO_2 in the atmosphere. Was it really on the increase, and had Callendar been right all along? In the late 1950s an American scientist, Dave Keeling, began to make very accurate atmospheric CO_2 measurements. With one instrument on the Mauna Loa volcano in Hawaii and another in Antarctica – both places far from human or natural influences that could distort the readings – he was able to measure CO_2 with unprecedented accuracy. In 1960, he reported that CO_2 was 315 ppm of the atmosphere, and increasing every year.

Keeling also showed that the level of CO_2 varied slightly with the seasons. It fell during the northern hemisphere summer and increased over the winter, owing to plant activity. There's much more land north of the equator than in the south, so there is more plant life. The plants soak up CO_2 from the air as they grow and the level in the atmosphere falls, reaching a minimum every October (late autumn). Plants then stop growing over winter and some die, CO_2 returns to the atmosphere and the level increases, peaking in May (late spring). It's as if the planet were breathing in and out. But superimposed on this annual cycle is a clear upward trend. The steadily climbing graph of measurements from Mauna Loa – the longest continuous record we have – is now known as the Keeling curve.

New Zealand began its own series of CO_2 measurements at Baring Head, a lighthouse station near Wellington, in 1970. Those readings closely match the Keeling curve but, being recorded in the southern hemisphere, don't show such a pronounced seasonal cycle. At the time of writing, the CO_2 level was a little over 380 ppm. In the mid-19th century it was 280 ppm (measured very precisely in bubbles trapped in Antarctic and Greenland ice). It has therefore risen by 36 percent over the last 150 years (see Fig 5, p. 98).

Before the Second World War, Callendar had sought to show a link between CO_2 levels and increasing temperature. After the war, global temperature stopped increasing. By the early 1970s, there was even some concern that the globe was cooling. New evidence from studies of ice cores and cores taken from sea and lake floor muds suggested that the climate could swing between ice ages and warm periods in what seemed like a short space of time – a few hundred years, perhaps even a few decades. Some speculated that the cooling since the war might mean we were on the verge of a new ice age. On the other hand, the role of CO_2 and other greenhouse-gases in warming the planet was much better understood. Computers had revolutionised weather forecasting by making it possible to use complex mathematical models to describe the way the atmosphere moved, so forecasts could stretch out several days ahead. It also became possible to model the movement and absorption of solar radiation in the atmosphere – a very complicated process. Putting the two together enabled the first real attempts to model the global climate system. By the late 1970s, the first global circulation models – attempts to build a description of the way the world's weather and climate system worked – were running on the supercomputers of the day. (Today anyone can download a much more sophisticated model and run it on a home computer.) And after 30 years without much movement, the global average temperature began to increase.

In 1979, the fledgling climate-science community held the first World Climate Conference, in Geneva. Papers reviewing the state of climate science showed there was a "clear possibility" that increased CO_2 levels could cause climate change. The data continued to improve, thanks to increasingly powerful computers and the discovery, from more ice cores, that CO_2 levels had been much lower during recent ice ages. By 1985, a group of experts meeting in Austria was confident enough to announce that "in the first half of the next century, a rise of global mean temperature could occur which is greater than any in man's history". That meeting also called on the world's governments to take action to address the dangers.

As a result, in 1988 the World Meteorological Organization and the United Nations Environment Programme set up the IPCC, a group of government representatives and the world's leading climate experts to provide the best possible advice on climate issues. The IPCC's job was to come up with a consensus picture of the state of knowledge and a summary of the latest research that governments could use to inform policy making. The IPCC's mission was complicated by the fact that the final report also had to be endorsed by a group of government delegates representing a huge range of national interests, from small Pacific nations fearful of sea-level rises to oil-producers such as Saudi Arabia, and big energy-users such as the USA and China. When the IPCC was set up, this structure was expressly intended to ensure that the reports were conservative. Getting scientists and government representatives to agree on the precise wording of the reports would ensure that they wouldn't prematurely rush to judgment or overestimate the certainty of findings.

The first IPCC report, in 1990, stated that the world had been warming but that it was too early to be sure whether CO_2 emissions from human activities were the cause. The IPCC predicted that it would take another 10 years to be sure, but it also projected that by the middle of the next century when CO_2 levels would be twice the 19th century (pre-industrial) level the global average temperature would increase by 1.5–4.5°C. The report was presented at the second World Climate Conference, held that year.

The next report, in 1995, suggested the same temperature increase by the middle of the 21st century. The third report (2001) found that most of the observed warming of the preceding 50 years was "likely to have been due to the increase of greenhouse-gas emissions". The fourth report (2007) has further underlined the point. The world is getting warmer, and greenhouse-gas increases caused by human activities are the cause. Unless we take steps to reduce those emissions and stop the increase in greenhouse-gases in the atmosphere, we will soon be living in a different world.

2

The climate system

Climate is average weather: all the weather we get, averaged over a period of time so that we end up with numbers like those for "January rainfall in Waipara" or the "average daily maximum temperature in Auckland in August". Mark Twain put it very well: "Climate is what we expect; weather is what we get." It's also about how often weather events happen – the frequency of damaging wind or rain, floods or frosts. By looking at how often these things happen, we're able to suggest that a massive rainstorm might happen once in, say, 20 years. Really big events might be said to be 100-year events (or longer), implying that on average they happen once in 100 years.

Weather is solar powered. The sun heats the sea and the air, setting them in motion. The oceans covering 70 percent of the planet's surface move that absorbed heat around the globe. As the water warms, some of it evaporates to become water vapour – the most important of the greenhouse-gases. The mixture of air and water vapour moves heat around in the atmosphere. It takes heat to make water change from liquid to gas, so as it evaporates off the sea it cools the remaining water, just as sweat evaporating from your skin cools your body. When water vapour condenses back to a liquid it gives up heat, and it gives up even more heat if it freezes.

The Earth's rotation creates great swirls of ocean and atmosphere that move around the globe from west to east in high latitudes. Most of the energy from the sun is absorbed between about 20°N and 20°S. (Further north and south the total surface area is much smaller, so far less energy is received). The Earth's climate system – the atmosphere and oceans working together – moves that heat towards the poles, where the sun's power is greatly reduced and in winter, absent. Hot, wet air rises in the tropics, moves towards the poles, cools and falls to the surface, creating large areas of high pressure that move around the globe just north and south of the tropics. Near the poles, cold air sinks and spreads out, is heated by the oceans as it moves towards the equator and then rises again, creating another circulation cell. Between the two cells of overturning air, great bands of westerly winds circle the planet, broken by the endless series of low-pressure systems that drive so much of the weather in the mid-latitudes. All this movement takes place in the bottom layer of the atmosphere, the troposphere. It's about 17 km thick at the equator but only 7 km at the poles. In the troposphere the temperature of the air falls with altitude – roughly 6.5°C for every 1 km, which is why it's cold at the top of Aoraki/Mt Cook. Above the troposphere is the stratosphere, and here temperature starts increasing again. The boundary between the two is called the tropopause.

This global circulation is what weather forecasters and climate scientists model in their supercomputers. Forecasters take measurements of the atmosphere such as pressure, humidity and temperature, at different heights for lots of points in the area of interest – basically a big three-dimensional grid pattern – and then calculate what's going to happen at each point as time passes. Each point is influenced by conditions at the points around it, by the sea or land (including mountains) underneath it, by the time of day and by the season. The accuracy of the resulting weather forecast depends on how good the initial measurements are, and how far into the future the forecasters are trying to predict. Forecasts for up to four days are generally very good, but after that the accuracy declines. Any errors in the initial data end up being multiplied to the point that the forecast is useless. It's called an "initial conditions" problem. The better you can measure the starting conditions, the better the forecast. At the same time, the smaller the grid you use for your computations (meaning the more data points and computing power you have), the better you can reflect the real weather as it interacts with the land and seascape. Major forecasting centres around the world produce global forecasts, while national and local forecasters use much finer grids and more detailed representations of the shape of the land to derive the forecasts we use every day.

For longer-range forecasts, to get around the initial conditions problem, forecasters use a technique called a Monte Carlo simulation, named after the famous casino. It's a way of determining the odds of a particular forecast being accurate. They run their weather model many times, incorporating many small, random changes in the initial conditions. Often the small changes make little difference and the forecast results all look similar, so the forecasters can be confident that the weather is inherently predictable. Sometimes the models diverge rapidly, and long-range forecasting becomes very difficult. Using these methods, the NZ Met Service is able to offer visitors to the rural weather section of its website a regional forecast out to 10 days. These forecasts are not hugely detailed, and there's no guarantee that they'll be right at 10 days, but they are still useful to farmers and growers. More computer power and better measurements of the initial conditions might extend forecasts out to two weeks, but for the time-being nobody expects to do much better than that.

Modelling the climate is weather forecasting on steroids. The global circulation models (GCMs) employed by climate scientists use the same computer code to represent the behaviour of weather systems as that in meteorologists' models. They have to generate realistic weather patterns inside the model so that the climate, the average of weather, can be worked out. But the GCMs are not attempting to produce weather forecasts. They're interested in the statistics derived from lots of weather events. The virtual weather inside the models looks like the real thing, but isn't a meaningful forecast. Instead of the initial conditions being the problem, GCMs have to take into account the things that are driving climate change – heating or cooling the planet. These are known as the "forcings", and include changes in the amounts of greenhouse-gases, changes in the albedo of the Earth as ice sheets vary in size or trees grow or are cut down, the effects of industrial pollution (sometimes warming, sometimes cooling), changes in the output of the sun, the cooling effects of dust, and aerosols from volcanic eruptions. These are the drivers of climate change – the factors that change energy flows in the climate system. The models also have to incorporate feedbacks. These are systems that respond to changes in forcings by amplifying those changes (a positive feedback) or damping them (a negative feedback). One of these feedbacks is the amount of water vapour in the atmosphere. As the atmosphere warms, the amount of water vapour it can carry increases, and because water vapour is a greenhouse-gas, it adds to the warming.

GCMs use large-scale three-dimensional grids that cover the entire globe and divide the atmosphere into layers in each cell of the grid. It's not just a matter of modelling the atmosphere: another important part is representing the ocean underneath the atmosphere. The amount of heat moved around by ocean currents is huge, and has a very marked effect on weather and climate all over the globe. (Britain and western Europe would be much colder if the Atlantic weren't shipping large quantities of heat across the ocean in the Gulf Stream, and north into the Arctic via the North Atlantic Drift). In order to accurately simulate the climate system, an atmospheric model has to be run in association with an ocean circulation model – with the atmosphere "coupled" to the ocean – and include other models to represent sea ice and ice sheets. Some GCMs also include biosphere or carbon-cycle models, so they can see how climate changes interact with those systems. A dozen or more of these complex models are run at institutions around the world, with some of the best known being at GISS in the USA and the Hadley Centre in UK. In New Zealand, the National Institute for Water and Atmospheric Research Ltd (NIWA) uses the Hadley Centre model to drive its own local modelling of climate change.

These models are, as you might expect, immensely complex and have taken decades to develop and refine. They demand huge computing resources, because climate modellers have to perform many runs examining changes in forcings. The steady increase of computing power also makes it possible for the models to become more complex and to use ever-smaller grids and layers, and therefore get closer to emulating what's happening in the real world.

Climate modellers have two main ways of testing how good their models are. The first is to see how well they represent the way the climate system behaves today. Run with all the forcings set to current values, do they produce the same rainfall and temperature patterns we see in the real world? For the most part, they do. One of the key tests is to look at how the models respond to large volcanic eruptions. When Mt Pinatubo in Indonesia erupted in 1991, it threw a huge amount of sulphur particles (called sulphate aerosols) high into the atmosphere.

These aerosols blocked some of the sun's incoming radiation and over the next 18 months cooled the whole planet by 0.5°C. Because we have accurate, detailed data for the eruption and its effects on the climate, we can put the forcings agreement (the aerosols) into the models and see if they produce the same results as were actually observed. They do.

The second kind of test is to take what we know of the change in forcings over the last 100 years and run the model to see if it can produce the same patterns of climate change and global temperature increase. This is a "hindcast" or forecasting backwards. Taking this approach a step further, we can feed into the model the forcings estimated for much earlier times – we have a good idea about greenhouse-gas levels and ice-sheet size during ice ages, for instance – and see what we get. This can provide information about climate patterns during the ice age, but also sets limits on various parameters in the models themselves. The models are said to be constrained by matching what we know of the ice-age climate, because if they don't produce the same general results then they can't be working properly. If the models pass those tests – and all do within reasonable limits – then they can make useful suggestions about what might happen in the future.

A good example of this comes from testimony that NASA climate modeller James Hansen gave to the US Senate in June 1988. Hansen's message was that global warming was likely to have an impact in the near future, and to demonstrate his point he showed a graph of global temperatures projected for the next few decades by the GISS model. Allowing for increases in CO_2 and other greenhouse-gases, and even for the cooling effects of large volcanic eruptions, the model predicted a steady rise in temperature. Nearly 20 years on, the scenario he labelled "most likely" has turned out to be pretty much spot-on.

GCMs model the movement of heat through the climate system, and so have to take into account the way water moves through the entire hydrological cycle – evaporating from the surface of oceans and lakes, condensing into clouds, and then raining or snowing into the sea, on to land, into rivers and back to the sea. They also have to take into account the carbon cycle, which is not only important in working out what might happen to the global climate but essential if we're attempting to work out how to try to keep warming within manageable bounds.

There's a lot of carbon around. The CO_2 in the atmosphere is equivalent to 750 billion tonnes (750 Gt C) of carbon (you convert weight of CO_2 to weight of carbon by dividing by 3.66). Carbon dioxide dissolves in water readily, especially cold water. Atmospheric CO_2 is mopped up by the oceans, and the relatively warm surface layers hold 1020 Gt C. That's dwarfed by the amount dissolved in the much colder deep oceans, estimated at 38,100 Gt C. Vertical movements of water in the ocean, downwelling and upwelling, can move significant amounts of carbon into and out of the system. Warming oceans will not only absorb less CO_2, but they will return some to the atmosphere – a positive feedback. Around 100 Gt C move between ocean and atmosphere every year, and at the moment the oceans are absorbing a bit more (around 2 Gt C) than they're releasing. The cold Southern Ocean circulating around Antarctica is a major site of that absorption.

The biosphere, all the Earth's living things, contains another 1900 Gt C. Just as water moves through the hydrological cycle, so carbon moves around in the carbon cycle. Plants breathe in CO_2 and use it to build the organic molecules that make up their bodies. When they die, some of that carbon returns to the atmosphere as methane, which breaks down over time to CO_2. If conditions are right, the plant material might turn into peat or, over geological time, coal or oil and lay down new stores of fossil carbon. In the oceans, calcium carbonate in the shells of microscopic sea creatures can build up on the sea floor and over millions of years turn into limestone and chalk. In very rough terms, the movements of carbon throughout the system are in balance although, as we saw when looking at the CO_2 levels in the atmosphere during the ice ages, over long periods of time the balance can shift as forests grow or die, and as the ocean warms or cools (see Fig 6, p. 100).

Total emissions of CO_2 caused by human activity accounted for 7.9 Gt C in 2005. One part per million of CO_2 in the atmosphere is equivalent to 2.13 Gt C, so if all that had stayed in the atmosphere the CO_2 level would have risen by 3.7 ppm. But the actual rise was only 2.42 ppm (equivalent to 5.2 Gt C), so "carbon sinks" are doing us a favour at the moment by absorbing 2.7 Gt C per year of the carbon we emit (see Fig 7, p. 99).

Since the atmospheric CO_2 level started climbing in the 19th century, it's risen from 280–380 ppm. Over that time, we've emitted over 300 Gt C, but only 213 Gt C has ended up in the atmosphere. Natural carbon sinks, the sea, land (especially soil) and plant growth, have stored up about a third of the CO_2 output and reduced the potential atmospheric increase by more than 40 ppm.

About three-quarters of man-made carbon emissions come from burning fossil fuels – oil, gas and coal – and the manufacture of cement. The rest comes from changes in land use, especially cutting down or burning forests. The total amount of fossil fuel underground is probably around 6000 Gt C, most of it as coal. Oil and gas combined probably account for about 1000 Gt C if we allow for reserves not yet discovered. Estimates of the amount of fossil carbon available for extraction vary considerably, and are in some cases controversial. The amount of oil remaining depends on who you talk to. Oil companies and oil-producing countries tend to be more optimistic than analysts, who suggest that we're approaching (or have passed) peak oil production. But one thing is clear: there's still enough readily available oil, gas and coal to boost atmospheric CO_2 to more than twice the pre-industrial level.

Methane is another important player in the carbon cycle, partly because it is a highly efficient greenhouse-gas – molecule for molecule 23 times as effective as CO_2 at absorbing radiation. Over the last 200 years, the methane concentration has increased from around 700 parts per billion (ppb) to 1750 ppb. Natural sources of methane include wetlands and termite mounds, the oceans, and gas released by undersea methane clathrates (naturally occurring methane/water mixtures that look a bit like ice, but which fizz and can be set on fire). There are large deposits of clathrates off the east coast of the North Island. We've added to that methane by directly releasing it from oil fields and leaking pipelines, from burping (not farting) cows and sheep, from rotting materials in landfills, burning biomass, and especially from rice agriculture – artificial wetlands on a large scale. Some methane is absorbed by soil, but most is oxidised over time – about 10 years – into CO_2. In recent years, the steep growth in the methane concentration has begun to level off, perhaps owing to a slowdown in the expansion of rice agriculture or changes in growing methods, but perhaps also because of better pipeline maintenance.

A lot of carbon is locked away in clathrates – possibly as much as 2500 Gt C – and there is evidence to suggest that ocean warming could cause them to melt, creating a large release of methane. This is believed to have happened at least once, 55 million years ago, when the climate record suggests there was a sudden injection of large amounts of methane which caused an ecologically disruptive warming. There's also a lot of methane locked away in the permafrost landscapes that circle the North Pole – the equivalent of perhaps as much as 400 Gt C. The permafrost is already showing signs of large-scale melting and could be releasing significant quantities of gas. It's a classic case of positive feedback: more warming melts more permafrost, releasing more methane and causing further warming. Given the extent of recent climate change in the Arctic, this is a serious issue and a matter for urgent research.

Clearly, there are many difficulties associated with modelling the global carbon cycle. Although we know a great deal about life on our planet, there's still much we don't know, particularly about oceanic ecosystems. There are surprises elsewhere, too. It was only recently discovered that living plants can emit methane. How much, and how significant this is for the methane part of the carbon cycle, is not yet known.

Despite the uncertainties, researchers are linking carbon-cycle models to GCMs to try to understand how the fluxes in the carbon cycle will change as the climate warms up. In the same way that you have to couple the atmosphere to the ocean to get a good global circulation model, so you have to couple the carbon-cycle model to the climate model to see how the feedbacks and climate interact. Work at the Hadley Centre has shown that those feedbacks can be very significant. As the climate warms, soils which presently act as carbon sinks begin to release more carbon than they absorb. The Amazon rainforest is thought to be very vulnerable to increased drought as the climate warms, and if it dies and turns into savannah it will release a great deal more carbon into the atmosphere. Add in the effects of methane emitted from permafrost, and you have feedbacks that could put the equivalent of 100–200 ppm of CO_2 into the atmosphere over the next century – enough to give the warming process an enormous boost. This work is still in its early days, but it suggests that if we are to reduce global warming we will have to monitor the carbon feedbacks closely and minimise them. It's worth noting that the current IPCC projections do not include carbon-cycle feedbacks, and so may understate our real vulnerability.

In the next chapter we're going to see what the latest research suggests the climate has in store for us, but before we do that there are a couple of important concepts to understand: the climate commitment, and the use of scenarios to produce projections. Climate commitment refers to the future warming that will take place as the climate system catches up with the impact of all the greenhouse-gases we've added to the atmosphere. At the moment, the Earth is absorbing more energy than it's radiating into space – about 0.85 Wm^{-2}. To get back into balance, the planet's temperature has to increase until it can radiate away the same amount of energy as arrives from the sun. The present energy excess has to go somewhere, and it's mostly going into the oceans. They can store an enormous amount of heat, and so take a long time to warm up. To get the planetary temperature to rise therefore takes a lot of energy. If we could hold greenhouse-gases at current levels, the temperature would rise another 0.6°C and take around 30 years to get there. That's the same as the temperature increase we've seen since the late 1970s, and at roughly the same rate: 0.2°C per decade. That's something we can't prevent, and it's very important to remember when making climate-change policy.

Related to the issue of climate commitment are the concepts of equilibrium and transient climate responses. The equilibrium response is what you get when the planet has a chance to get back into energy balance. In modelling terms, this means choosing a future level of greenhouse-gases (the figure of twice the pre-industrial level is often used), and then looking at the climate when it's had a chance to stabilise.

The transient climate response is what you see when you run a model with constantly changing forcings – which is what we're doing in real life. If the world decided on aggressive action to cut carbon emissions tomorrow, it would take years or decades to put all the required changes in place, and so the amounts of atmospheric carbon would continue to climb for some time before stabilising. This is called the emissions trajectory – if you want to have a look at the climate in the 2050s, you feed in an emissions trajectory and look at the transient response for that time. If 2050 is when you think you can stabilise greenhouse-gas levels, then because of the climate commitment you still won't get back to an equilibrium climate until close to the end of the century. There can be important climate differences between equilibrium and transient responses. In New Zealand, for instance, the equilibrium response to warming suggests that the strength of the westerlies that bring rain to the West Coast and snow to the Alps may be reduced, resulting in less precipitation, whereas transient model runs suggest they will be stronger and the West Coast therefore wetter. An important difference for Coasters – and for the rest of us, as we'll see later.

Climate models do not produce climate forecasts the way that weather forecasting predicts the weather. Given good information about the initial conditions, and a good model, meteorologists make good forecasts about weather events. Climate models can't do the same thing because how they behave depends on how the forcings and feedbacks change. They also have natural variability built in, reflecting the way that the real climate varies from year to year. That's fine if you want to know about the statistical probability of weather conditions in the future, but not much use if you want to know whether spring 2012 in Hokitika will be warm, cold or just average.

The other major limiting factor in how GCMs perform is all about predicting human behaviour – how will future generations manage climate change and adapt to control the causes? We have good information about how climate forcings – the things that drive climate – have changed over the recent past, but have to rely on guesswork for the future. At its simplest, this could be a straight-line extrapolation from where we are now: 381 ppm CO_2 in 2007, increasing at 2.5 ppm per year, puts us at 406 ppm in 10 years and 431 ppm in 20 years. Feed those numbers into the climate model and you can see what they suggest the climate of 2027 might be like. But tomorrow almost certainly won't be the same as today. We know that China and India are industrialising fast, building lots of CO_2-emitting power stations. At the same time, forests are being cleared in Southeast Asia and the Amazon. Perhaps the rate of CO_2 growth will increase. But what if we start applying clean-coal technology, using more renewable energy resources and improving the fuel efficiency of our transport fleets? Perhaps the rate of CO_2 growth will decline. The numbers that are fed into the models depend on the answers we select for those sorts of questions. Climate scientists use a range of specially developed scenarios that are designed to provide plausible emissions trajectories, given a range of different population and economic growth projections. To the climate modeller the important thing is the emissions trajectory, not the policy decisions it reflects. The models provide us with a projection for what global and regional climates might be like for given levels of greenhouse-gases. But how we get there is up to us.

3

The state of science

The Intergovernmental Panel on Climate Change (IPCC) was founded in 1988. Its job is to pull together all the strands of climate research and provide governments with a snapshot of what the best available knowledge is telling us about climate change. It provides a consensus view – an assessment that's approved by an overwhelming majority of recognised experts and reviewed by all the participating governments – and sets the agenda for further research. This is no mean feat, given the range of scientific disciplines involved, from atmospheric physics and climate modelling to oceanography, glaciology and ecology.

The study of climate is a fast-moving field. Over the period that I researched and wrote this book, new work was being published in a steady and newsworthy stream. The latest IPCC report (the Fourth Assessment Report, or AR4) summarises the state of the science in 2005/6.

The IPCC operates through three main working groups. Working Group One (WG1) is concerned with the science of climate, Working Group Two (WG2) looks at the impacts of change and how to adapt to them, and Working Group Three (WG3) considers how to mitigate – that is, lessen or eliminate – the harmful effects of change by reducing the emissions that cause it. Each group produces its own report, and these are then pulled together into the "synthesis report". In addition, the IPCC produces occasional special reports on technical issues. The most recent special report considered CO_2 capture and storage, looking at how technologies might be developed to allow fossil fuels to be burnt without adding huge quantities of greenhouse-gas to the atmosphere.

WG1 is the first stage of the process. It's here that climate scientists describe our best understanding of what the climate system has in store for us. The report is, in effect, a textbook of our current knowledge. These findings are then fed into the work of the other two groups. What climate modellers tell us about probable global and regional changes is fed into WG2, so that group can consider what might happen. WG3 is concerned with how to reduce emissions, to limit the potential damage identified by WG2, and how policies and technologies might be deployed to reduce emissions. This book follows much the same pattern:

basic climate science first, then what it tells us about the likely future climate for New Zealand and the impact it will have, followed by a look at what we can do about it. Because of this setup, WG1 releases its final report first – the "summary for policymakers" section was released in early February 2007 – followed in April and May by the WG2 and WG3 reports, and the synthesis report before the end of the year. The IPCC cycle will then start again, with the fifth report due in 2013.

So what is the state of our knowledge of the climate system and where it may be heading? One parameter is the "climate sensitivity". This is defined as the equilibrium response of the climate system to a doubling of CO_2 above pre-industrial levels (from 275–550 ppm). That response includes temperature changes (and their distribution round the planet) and the effect on sea level. "Sensitivity" in this case specifically excludes feedback from the carbon cycle and relatively slow responses such as the melting of ice sheets, but includes faster feedbacks such as cloud, water vapour and sea-ice coverage. It's more or less what Arrhenius set out to calculate (see p. 22), and is appropriate when considering a timescale of a century or so. Over longer periods, large ice sheets have time to melt, and will cause bigger changes in sea level. The IPCC puts climate sensitivity at 3°C plus 1.5°C or minus 1°C in the global average temperature. In other words, double the pre-industrial CO_2 level and the temperature rise will top out in the range 2–4.5°C. This is partly derived from climate-model experiments, and confirmed by studies of how the climate has moved in and out of ice ages over the last million years. We know the ice age and interglacial climate forcings from a wide range of paleoclimate evidence – the gases in ice cores, the marks left by ice sheets as they melted – and we know that the global average temperature changed by 4–6°C between full ice age and warm interglacial. When you work out climate sensitivity using this information, it comes to 3°C (plus or minus a bit). This number hasn't changed much since the first IPCC report, but the level of confidence in it has increased.

Although we have a very good idea of the theoretical climate sensitivity, that's not the same thing as knowing how the climate will behave in the future. If we were still in a normal glacial/interglacial cycle, with the world's forests and other ecosystems intact and CO_2 still in its natural range (180–280 ppm), then the climate sensitivity we've calculated would still hold true. But we have added another 100 ppm CO_2, a lot of methane and all sorts of other atmospheric pollutants, and we've cut down a lot of trees and turned large areas of the Earth into agricultural land. Climate sensitivity as defined above is a useful measure of how well we understand the climate system, but it is not a projection or prediction about how the full climate system – carbon cycle and all – will react to a doubling of CO_2 above pre-industrial levels. That is much more complicated and uncertain.

As we saw in the last chapter, the amount of warming projected for any future period is determined by the climate system's transient response, because the climate is catching up with the full effects of the emissions trajectory we have chosen to follow. At the same time, different models give slightly different results – partly because they include natural climate variability, and partly because they are constructed differently. They might differ in the way they handle sea ice, or clouds, or in the type of ocean model they're coupled to. The IPCC therefore uses multiple runs of single models, as well as looking at the range of results across different models. When broad patterns emerge that all the models agree on, the IPCC regards their projections with high confidence. The first and most obvious of these results is that the different parts of the planet warm at different rates. By definition, a global temperature is an average across the whole planet. Hidden in any overall increase there will be places that warm less, and some that warm more, both around the Earth and up and down in the atmosphere.

All models project that the stratosphere will cool while the Earth's surface temperature increases. As the troposphere is trapping heat close to the Earth's surface, less heat radiates outwards to warm the stratosphere above. This is complicated by the role that ozone plays in warming up the stratosphere. Stratospheric ozone levels have been severely reduced over recent decades, owing to chlorofluorocarbon (CFC) emissions (from refrigerants and aerosol propellants, though these are being phased out under the Montreal Protocol of 1989). CFCs are most efficient at destroying ozone in very cold conditions (which is why we have the ozone hole over Antarctica), so the stratospheric cooling caused by greenhouse-gas increases is making things worse. The ozone-layer recovery is being significantly slowed down.

Another very confident prediction is that the poles will warm up more than the tropics. In part this is for the same reason that we expect nights to be warmer when there is more greenhouse-gas in the atmosphere: less heat will be radiated away. The poles lose a lot of heat to space during their long winters, but more greenhouse-gas means less heat loss and therefore warmer winters. At the same time, the planetary heat engine transfers heat from the tropics towards the poles, so the heat gained there is transferred northwards and southwards. The area of the poles is much less than that of the tropics, so the heat becomes concentrated in the polar regions, a phenomenon called polar amplification. In all warming projections, the Arctic warms much more quickly than the Antarctic. This is because the Arctic is an ocean surrounded by large continental land masses. Ocean currents can transport heat from warmer latitudes all the way under the sea ice to the North Pole. In contrast, the Antarctic is a large and very cold continent, surrounded by a ring of cold ocean currents. Although the southern oceans can (and will) warm it up eventually, they can't transfer heat all the way to the pole.

Land areas are expected to warm more than the sea, and to warm more in winter. The further you get from the sea, the less influence the ocean has on temperature, because the sea isn't smoothing things out by absorbing and releasing heat. This means that inland on the continents summers are warmer and winters cooler than in coastal regions. The closest thing New Zealand has to a continental climate is in Central Otago, where summers can be baking hot and winters bitterly cold. As the climate warms, the centres of large landmasses will have both warmer summers and warmer winters, and the temperature increases will be greater than average. The northern hemisphere has much more land than the south, and so polar amplification is itself enhanced by the land effect, making the northern hemisphere warm faster than the south.

With more heat going into the ocean there will be more evaporation.

This means more greenhouse warming from the extra water vapour, and more rain. It's predicted to increase in the tropics and at high latitudes, but to decrease in mid-latitudes. Places that are already dry could become drier. There is also a likelihood that when it rains it will rain harder. Even if the total rainfall in a region doesn't change much, fewer but heavier falls could cause more flooding and erosion problems.

Another consequence of heat going into the oceans is that sea level will rise. As water warms, it becomes less dense and expands. The sea level is currently rising by about 3 mm a year, and the rate may be increasing. Some of that is due to oceanic warming, but there's a growing contribution from melting ice. There's a lot of water stored in the world's glaciers and ice sheets. During the last ice age, around 20,000 years ago, a huge amount of water was piled up in the form of great ice sheets covering North America and northern Europe, as well as Greenland and Antarctica. As a result, the sea level was 120 m lower than it is today. New Zealand was one large island, from the Stewart peninsula up to far beyond Cape Reinga, and its glaciers were much larger. The snouts of the Fox and Franz Josef glaciers ran into the sea, well to the west of the present coastline, and in Canterbury the Rakaia glacier stretched out onto the Canterbury plains. As all that ice melted and the glaciers retreated, the sea level rose – sometimes quite rapidly. About 14,000 years ago there was a surge of melting and the sea level rose by about 25 m – perhaps in as little as 500 years. The sea reached roughly modern levels about 6000 years ago, and hasn't changed greatly since (see Fig 8, p. 99).

The ice left in the world's remaining mountain glaciers (not counting Greenland and Antarctica) would increase sea level by about half a metre if it all melted. There is much more ice covering Greenland and Antarctica: Greenland has enough for a 5–7 m rise, and the West Antarctic ice sheet could add the same amount. If all the rest of the Antarctic ice melted, the sea level would rise by a further 60 m. It's obvious that if things get warmer, more ice will melt, but there's another factor to consider. Snow falling in the middle of the ice sheets, or in the feeder basins of mountain glaciers, replaces some of the ice lost to melting. If the rate of snowfall, and therefore ice accumulation, exceeds the rate of melting, then the mass of ice increases. That's how an ice age starts. When melting wins, the ice age ends and the sea level rises. At the moment, it's clear that for most of the world's glaciers melting is winning. The annual loss of ice mass is around 300 km^3. Snowfall on the Greenland and Antarctic ice sheets has been estimated to be equivalent to about 8 mm of sea level every year, falling on the high cold inland areas. Nevertheless, melting around the edges is more than enough to offset that accumulation. Until recently, it was thought that increasing snowfall sufficient to compensate for melting at the edges of the sheets, but the latest studies suggest that Greenland could be losing more than 200 km^3 of ice per year, and the Antarctic as a whole more than 150 km^3. It's too early to tell if these numbers are high or low, or if this is a long-term trend. The impact this will have on sea level is therefore not clear. The IPCC takes a conservative stance, suggesting that sea levels will rise by between 18 and 59 cm owing to thermal expansion and meltwater from glaciers, including a contribution from Greenland and Antarctica at the rates measured in 1993–2003. They do warn, however, that if melting from those ice sheets increases linearly with projected temperatures, that could add 10–20 cm to the sea level by 2100, and that larger amounts can't be excluded. This is one area where the research over the next few years is going to be fascinating to watch.

The traditional view has been that ice sheets, being big, cold and high, will take thousands of years to melt – like giant ice cubes, just melting at the edges – but there is worrying evidence, particularly from Greenland, that it may be a much more dynamic process. During summer in Greenland, on the lower parts of the ice sheet, lakes of meltwater form on the surface and then plunge down sinkholes (called moulins) to the underlying rocks hundreds of metres below. The water warms and lubricates the base of the ice sheet, helping it to flow over the rock towards the coast. At the same time, the huge glaciers that feed the ice into the sea have been speeding up. The largest glacier in Greenland, the Jakobshavn Isbrae, is the outlet for 6.5 percent of the total ice sheet. Between 1997 and 2003 it more than doubled its speed, from 5.7–12.6

km per year, and its calving front, where the ice breaks off into the sea to form icebergs, has been retreating rapidly. The area of Greenland over which surface melting occurs every summer has increased significantly. Similar things are happening on the Antarctic Peninsula. What this means for sea levels is not yet clear, but researchers are working hard to improve the models they use to describe the way ice sheets move, and keeping a close eye on how the ice changes from year to year. The melting surge of 14,000 years ago was probably caused by an ice sheet collapsing as the last ice age ended. At that time the sea level may have risen by as much as a metre every 20 years – far faster than humans have experienced since the beginnings of civilisation. No one is suggesting (yet) that such rapid sea-level rise is likely in the near future, but it does give an uncomfortable hint of what may be possible in a rapidly warming world.

Because GCMs use relatively coarse grids to map the surface of the planet, they can provide only a rough idea of what happens on a regional scale. The best they can offer are generalised statements such as "warming will exceed the global mean by 40 percent in central Asia and Tibet in winter". To get down to the much finer scale required to describe what will happen locally, it's necessary to take what the GCMs say about the regional climate, and translate that into useful information such as the annual rainfall at Bluff. Local climate depends on all the same factors that affect local weather – the shape of the land, its relationship to the wind and sea, and so on. Taking the large-scale projection and working out the relationship to regional climate is called downscaling, and there are two main approaches – statistical downscaling and dynamic downscaling.

Statistical downscaling works by examining the present climate at various places around the country and working out how key climate features such as rainfall and temperature alter as the averages predicted by the GCMs change. This assumes that the underlying probability of a weather event – say, a summer 1°C hotter than average – doesn't change when the average changes. If a hot summer is a one-in-10-year event when the average summer temperature is 20°C, statistical downscaling assumes that it will still be a one-in-10 chance when the average is 21°C. This assumption is not necessarily true. For instance, the increase in water vapour and energy in the atmosphere is expected to cause more intense rainfall, so assuming that the patterns of heavy falls will change smoothly may be wrong. The big advantage of statistical downscaling is that it doesn't need huge amounts of computer power to produce useful results. Most of the work NIWA has done on regional climate in New Zealand has used this technique.

Dynamical downscaling takes information from a GCM and uses it to drive a regional climate model (RCM). The RCM uses a much smaller grid than the GCM, and provides a much more detailed representation of how land, sea and atmosphere interact. It's exactly the same sort of relationship that exists between the big global weather forecasting models and the local weather models that provide the forecasts we use every day. The GCM provides the big picture, the RCM computes the detail. This approach is much more demanding on computer time, requiring a GCM and RCM to run in parallel, but it can provide information missing from statistical downscaling. Early work with NIWA's regional climate model, for instance, suggests that rain events heavy enough to cause damaging floods could become significantly more common in certain parts of the country by the end of the century, and that the big temperature rises will be experienced in the Southern Alps.

As we saw in Chapter 2, to get the models to give us a realistic idea of what the future climate might be like, we have to feed them with assumptions about how we think the amounts of CO_2, methane, nitrous oxide and other greenhouse-gases in the atmosphere are going to change over the rest of the century. We also have to look at how other forcings such as air pollution might change (for better or worse) and project how all these factors will interact. This involves asking all sorts of questions about the future and choosing a range of plausible answers. The IPCC has created a range of 40 scenarios to make a whole lot of emissions predictions and thus project a range of possible greenhouse-gas levels for the year 2100. For the AR4, they asked climate modellers to focus on six scenarios, which have total greenhouse-gas forcings in the year

2100 ranging from 600–1550 ppm CO_2 equivalent (CO_2e). This is a way of describing the effect of the methane, nitrous oxide and halocarbons in the atmosphere by expressing them in terms of the amount of CO_2 required to have the same warming effect. At the time of writing, the total greenhouse-gas forcings were 455 CO_2e, including about 383 ppm CO_2, 1745 ppb methane, 314 ppb nitrous oxide and smaller but still significant amounts of halocarbons. In other words, if there were no methane and other gases in the atmosphere, we'd need an extra 67 ppm CO_2 to cause the same amount of greenhouse warming. CO_2e is the standard measure used for emissions trading.

The scenarios used in the Third and Fourth IPCC reports were outlined in the IPCC's Special Report on Emissions Scenarios (SRES), published in 2000. They're built around four storylines, rather boringly called A1, A2, B1 and B2, each with a different projected outcome. Each is an attempt to create a plausible future history for human civilisation, and to suggest how combinations of population and economic growth around the planet might affect the factors that drive our climate. It's an interesting challenge. To put it in context, imagine someone conducting this sort of exercise at the turn of the last century. In 1900, cars were still rare and cities thronged with horse-drawn vehicles. Who would have predicted that the internal-combustion engine would soon dominate transport, or that aeroplanes would make cheap fast international travel possible? Could anyone have predicted the flow-on effects all this would have on international trade? What the world will be like in 2100 is just as difficult to predict. You might be able to have a reasonable stab at the next 10 years, but further out than that predictions become more like the stuff of science fiction. Instead the SRES attempts to paint a broad-brush picture of how we might *get* to the end of this century – a tool to illuminate the *choices we face* rather than making serious predictions. By covering a wide range of possibilities they generate a broad range of emissions trajectories for modellers to work with.

All of these scenarios expressly exclude any direct action to reduce greenhouse-gas emissions – in other words, they assume that the world doesn't attempt to meet Kyoto or any other set of emissions targets. They are also supposed to be equally plausible – whatever the amount of fossil fuels consumed might be, or the extent of forest suggested for 2100. The assumptions underlying the scenarios have been criticised by some, but that rather misses the point. It's the emissions that are important, not the storyline underlying the trajectory. However, designing policies to try to limit damage from climate change does require modelling to take account of emission-reduction targets, as we shall see when we look at mitigation (Chapter 9).

For the AR4, the six key scenarios were run through a wide range of models many times to generate best estimates of what global temperature might be at the end of the century (see Fig 9, p. 99).

At the low end, the B1 scenario has global average temperature increasing by 1.8°C compared with the average for 1980–1999 (with a range from 1.1–2.9°C), while A1F1, the most fossil-fuel intensive scenario, produces a best estimate of 4.0°C (with a range from 2.4–6.4°C). The full range, from the lowest figure for the lowest emissions scenario to the highest for A1F1, is therefore 1.1–6.4°C. That comfortably spans the full range from modest to disastrous, though it has to be remembered that these are transient responses and warming will continue beyond 2100 for decades, probably centuries, to come.

Let's put those numbers into context. Between 1850 and 2005, average global temperatures increased by 0.8°C. The best estimate for the next 100 years, if we keep our greenhouse-gas emissions at B1 levels, is a further 1.8°C increase – three times as much as we've experienced already. Although 1.8°C doesn't seem like a big number, it's a global average and in polar regions and over continents the change will be much larger than that. Think of it this way: the average temperature in October in Christchurch is 11.7°C. If you add 1.8°C to that, you have the average temperature for November. It's like adding a month to the beginning of summer and a month to the end. Not only will summers be hotter, they'll be longer, and winters will be warmer and shorter. Throw more emissions into the pot and the effect becomes even more pronounced. Once again, small numbers can have big effects. It's worth noting that even the low-end IPCC projection suggests the world in 2100 will be 2.6°C warmer than it was in 1850.

Modellers have also produced an estimate of what the global temperature rise would be if we had somehow limited greenhouse-gases to year-2000 levels. In that case, there would still be warming of 0.6°C over the century – about the same as the world has experienced in the last 30 years. In reality it is already too late to avoid a larger temperature increase than that.

There is another chilling conclusion to be drawn from the temperature-change data. The difference in global average temperature between the middle of an ice age and the warmest interglacial period is around 4–6°C. The planet's climate system can move from one extreme to the other in as little as 5000 years, which is about one degree celsius per 1000 years. That's regarded as fast – but our current rate of warming is 2°C in just 100 years. That's 20 times faster than the fastest changes we can find in the climate record. There's danger not only in the *amount* of warming, but in its unprecedented *speed*.

Whatever scenario you run through the modelling, the warming we will experience over the next 20–30 years is likely to be pretty much the same: around 0.2°C per decade. This is because much of the future warming is already inevitable (climate commitment) and the emissions trajectories are all similar for that period. It means we have a good idea about what's in store for a large part of our lifetimes: continued rapid warming, whatever we do. The only thing that could change that figure much would be a major volcanic eruption that threw up enough dust and aerosols to cool things down. But even that would give us only a few years' respite.

The pattern of warming around the globe – greatest over the poles, especially the Arctic, and over the continents, and smallest over the Southern Ocean and parts of the North Atlantic – shows up in all model runs and at all emission levels. Winter snow cover will decrease, and permafrost around the Arctic will continue to melt. Sea ice in the Arctic and Antarctic is projected to diminish, and the Arctic may even become ice-free in summer well before the end of the century. Heat waves and heavy rain events are very likely (over 90% probability) to become more common. Tropical storms are likely to increase in intensity, though perhaps not in number. The tracks of other storms are likely to move towards the poles, changing wind, temperature and rainfall over large areas. There will be more precipitation in high latitudes and in the tropics, but rainfall will decrease over the subtropics. There will be significant rainfall reductions over the Mediterranean and North Africa in winter, and over the Mediterranean and western Europe in summer. Brazil, Southern Africa and Western Australia will be dry in winter. The Caribbean and southwestern USA will be drier in both winter and summer.

Beyond the next 100 years, the IPCC is confident that warming and sea-level rise will continue for centuries, even if we manage to stabilise the level of greenhouse-gases in the atmosphere. Thermal expansion of the oceans, as heat is transported down to the deep ocean, could add as much as 0.8 m to the sea level by the year 2300, not including any melting of the Greenland and Antarctic ice sheets. A useful hint of what might be in store is that during the last warm interglacial, 125,000 years ago, global temperatures were around 1.5°C warmer than at present, and the sea level was 4–6 m higher.

There's another nasty consequence of pumping more CO_2 into the atmosphere: as we saw in the last chapter, CO_2 is easily absorbed by cold seawater and increases its acidity. Seawater is naturally slightly alkaline (8.1 on the pH scale). In pre-industrial times, its pH was about 8.2 (slightly more alkaline). That doesn't sound like a big difference until you realize that pH is a logarithmic scale, so that a drop of 0.1 actually represents a 30 percent increase in acidity. Over the next century, the surface of the ocean will go on absorbing a lot of our CO_2 emissions, becoming more acidic. Its pH will increase by between 0.14 and 0.35. This is simple chemistry, but its impact on ocean ecosystems is complex, and possibly highly damaging. Many oceanic organisms – corals, plankton, shellfish and micro-organisms use the carbonate in the sea water to build their shells, and those shells dissolve more easily as acidity increases. Research on the impacts this could have on life in the oceans is in its early days, but there is the potential for massive damage.

When oceanic acidity increased dramatically 55 million years ago the result was mass extinction of sea creatures with calcareous shells. If we needed another reason to cut back on burning fossil fuel, maintaining the health of the oceans would be a pretty good one.

4

The outlook for New Zealand

New Zealand is out on its own at the edge of the great Southern Ocean. Fly west from the bottom of the South Island and you'll travel two-thirds of the way round the world over empty ocean before you reach land in Patagonia. Fly east, and you'll eventually hit Chile. That bit of South America is pretty narrow. Our part of the world is mostly ocean, and our climate is dominated by it. To the south of the country, great belts of westerly winds circle the planet, helping to drive the cold Antarctic Circumpolar Current, the largest oceanic current on Earth. To the north of the North Island, the subtropical ocean is warm. The contrast between the cool south and warm north drives much of New Zealand's weather. It also helps to shape the way that we will experience global warming.

The amount of warming projected by global circulation models is usually expressed as a global average. Averages, of course, don't tell the whole story. As we saw earlier, some parts of the world, notably the northern polar regions and the large continental land masses, will warm more than most other places. New Zealand is certainly going to warm up, but the oceans around the country will slow the rate of warming we experience by acting like massive heat sinks. If the world heats up by 3°C on average by the end of the century, the northern polar regions will be more than 6°C warmer, while the continental interiors will be 4°C or more warmer. Down south, large parts of the southern ocean may warm by only 2°C. The sea will be our air-conditioner, keeping us cool – at least in comparison with the rest of the planet. Since 1950, for example, New Zealand has warmed by 0.4°C but the global average has risen by 0.6°C in just the last 30 years.

The New Zealand climate is also notably variable. Years can swing from being 1°C warmer than average to 1°C cooler under the influence of factors such as the El Niño/La Niña weather cycle and the movement of giant warm or cool bodies of water around the Southern Ocean. Although the country is expected to warm at the same rate as the global average, that variability will still bring us cool and warm years. Our advantage is that our warm years may be no worse than the cool years elsewhere.

Although global warming is expected to be at its most dramatic in the northern hemisphere, the southern polar regions will warm up too.

Polar amplification will be at work down south, and may have marked impacts on the edges of the continent. The Antarctic Peninsula, the bit of land that sticks up towards South America, has been the fastest-warming place in the world for the last 50 years, with temperatures increasing by over 2°C during a period when the global average increased by only 0.6°C. Ice shelves around the Peninsula have been breaking up, glaciers are moving faster, and grasses have begun to colonise the warmest areas. However, most of Antarctica is a high plateau, over 3000 metres above sea level in places – essentially a cold, dry desert because there is little precipitation, and with an average annual temperature of minus 37°C the ice there never melts. The oceans around Antarctica also help to keep it cool, by absorbing heat being shipped down from equatorial latitudes before it can get to the centre of the continent.

Having Antarctica and the cold oceans to the south and the warm sub-tropics to the north has one important consequence for New Zealand's weather and climate. The westerly winds (the Roaring Forties, Furious Fifties and Screaming Sixties) that blow round the planet and bring us much of our weather are expected to intensify. The extra heat being pumped into the atmosphere by global warming adds energy to the climate system, making everything more vigorous. For New Zealand, that means we'll get more westerlies, and that has important consequences for our regional climates, as we'll see later.

The sort of climate change we can expect in New Zealand has been modelled by the National Institute of Water and Atmospheric Research (NIWA), using statistical downscaling from the model projections contained in the IPCC's Third Report (2001). At the time of writing, NIWA was working to update these local projections based on the Fourth Report results, but the overall picture was expected to remain the same. NIWA is also beginning to use dynamical downscaling, which is good at capturing changes the statistical process can miss, such as changes in the frequency of extreme events.

Warming in New Zealand will match the global average for any scenario you may choose to examine – from a best estimate of 1.8°C by the 2090s (lowest-emission scenario) to as much as 4°C under highest-emissions scenarios. Over the next few decades, we are likely to see warming continue at 0.2°C per decade, though individual years will continue to vary from the overall upward trend. Warming will be greatest in winter, and there is a likelihood that the east and north of the country will warm slightly more than the rest. As the average temperature increases, the probability of extreme events – cold or hot – moves with it. Heat waves will become more frequent, and we can expect fewer frosts. That can already be seen in the climate data for recent decades: the number of cold nights and frosts per year has diminished by 10–20 since 1950 (see back flap).

The increase in the westerly flow of wind over New Zealand will have an important impact on regional rainfall. As the westerlies hit the west of both islands, they dump rain, leaving the east coasts of both islands drier and warmer. This pattern will intensify as the westerlies increase. By the end of the century, Taranaki, Manawatu, the West Coast, Otago and Southland will have higher average rainfall, while Hawke's Bay, Gisborne, and the eastern parts of Canterbury and Marlborough will be drier. Broadly speaking, the south and west of the country will get wetter, while the north and east will be drier. In particular, spring is expected to be drier in the north and east of the North Island. Large rainfall events are expected to become up to four times more common by the end of the century, especially in regions that see an increase in overall rainfall, and early projections based on dynamical downscaling suggest that rainfall will generally become more intense. Over the whole country, there could be 20 percent more rain on the three wettest days of the year, which on the West Coast translates to as much as 40 mm more rain on each of those days. In parts of Canterbury, there could be 50 percent more rain in the year's heaviest falls. The potential for flooding and erosion is obvious.

Drought is also expected to become more common as the climate changes. Drought is caused by lack of rain, and is made worse by heat and wind. Put all three together and you have a recipe for significant economic losses – the 1997–99 drought cost the New Zealand economy more than a billion dollars. Research by NIWA suggests that regions that are currently prone to drought could become even drier. By the 2030s, water-availability problems will intensify from Northland to the east coast of both islands. Under a low-to-medium-warming scenario, the risk of what is currently a 1-in-20-year drought might double by the 2080s in inland and northern Otago, eastern parts of Canterbury and Marlborough, and in parts of the Wairarapa, Bay of Plenty and Northland. Using a medium-to-high-warming scenario, the 1-in-20-year drought could become four times more likely in the 2080s in the east of the South Island from Otago to Marlborough, much of the Wairarapa, Bay of Plenty and Coromandel, most of Gisborne and a large chunk of Northland. In Whangarei, for instance, what is now a 1-in-20-year drought might happen every three years on average. Under that scenario, large parts of New Zealand would see damaging droughts happening twice as often as at present.

It's not only the risk of severe drought that increases. With eastern districts seeing less rainfall, the average moisture deficit (the difference between the water in soils actually available to plants and the amount they need for optimum growth) will increase. Soils could go into moisture deficit earlier in the growing season – a month earlier, under the high scenario – and the deficits could last longer into autumn. Under that high scenario, what we now think of as a medium-severity drought could be an almost annual occurrence by the end of the century. The implications for dryland agriculture are clear. Demand for irrigation water is going to increase, and pressure on groundwater and rivers will be considerable. Water is already a contentious issue in many parts of the country. Climate change looks certain to drive further controversy.

Land affected by prolonged drought is also particularly prone to erosion. If, as we've seen, rainfall events become more intense as the climate warms, then the potential for damaging run-off, gullying and soil loss could increase significantly.

With the general increase in westerly windflows comes a risk of increased strong and damaging winds, which could double in frequency towards the end of the century. There's no clear indication yet whether the frequency or intensity of storms will be affected, but it's possible that the number of tropical storms heading down towards us may decrease, although they may be slightly more damaging when they do arrive. The intensity of storms reaching us from the south and west may also increase, and their tracks could change, but once again it's not clear how this will play out.

A direct consequence of warmer – and shorter – winters is that the area of snow cover will be reduced. The permanent snow line in the mountains will rise, and the mountains will have snow cover for a shorter period of each year, but the amount of snow that falls may actually increase, owing to the intensification of rainfall events. Ski-field base stations may need to be moved upwards to reach the new snowline, but there may still be plenty of the white stuff at higher levels. This will also have a marked impact on New Zealand's glaciers. Over the last 100 years, the surface area of our glaciers has decreased by 35 percent, but since 1978 increased snowfall has offset the effect of warming, leaving things roughly in balance. As warming increases, however, that balance will shift, and further reductions of snow cover are likely. NIWA's latest studies suggest that by the end of the century warming in the Southern Alps could be significantly greater than over the rest of the country. Under one scenario, the Mt Cook region could have an average summer temperature increase of 6°C (from 14–20°C), which would raise the freezing level by a 1000 metres and have a marked impact on local glaciers and snow cover, and possibly cause large rockfalls and landslides.

One apparent paradox of global warming in the Alps is that it may cause some of our glaciers to get bigger. The famous West Coast glaciers at Fox and Franz Josef are fed with ice from large snowfields high up near the main divide. Exposed to the intensifying westerlies, they will receive increased snowfall and this will feed rapidly down to the glacier snouts. Fox and Franz Josef take only 5–8 years to respond to snowfall changes, so they will reflect greater snow accumulation relatively quickly. On the other side of the main divide, big glaciers like the Tasman respond much more slowly to changing snowfall at their sources. The Tasman, for example, is still shrinking after expanding

Fig. 11 The Tasman Glacier has lost ice and gained a lake in the last 30 years. The upper photograph shows the glacier in April 1973 (it looks black because it's covered in rock debris), already retreating from its Little Ice Age maximum. The bottom picture was taken in March 2007 and shows the large – and growing – lake littered with icebergs. *Trevor Chinn*

during the Little Ice Age. With the development of a lake at its tip in the 1970s, it is now fated to retreat over the next century as far back as the Ball Hut, leaving behind another large and deep lake. The fate of the other glaciers depends on what happens to precipitation in the basins that feed them – whether it continues to fall as snow or whether the warming climate causes it to become rain. Once warming is great enough to tip the balance from snow to rain for long periods of the year, the glaciers will be doomed. At least 25 percent of the total ice mass has disappeared since 1950.

Melting all New Zealand's ice will have no measurable impact on the global sea level, but we will feel the effects of the melting elsewhere. Since the middle of the 19th century the sea around New Zealand has risen by 25 cm, and since 1950 by 7 cm. Over the next 100 years it will increase by a further 18–59 cm, plus whatever extra results from the melting of the Greenland and Antarctica icesheets. However, the sea level at any given time is affected by many different factors – the state of the tide being the most obvious. Onshore winds and ocean currents can push the sea up against the land, and storms can cause a phenomenon called storm surge, where the sea level rises at the centre of a severe storm, owing to the low atmospheric pressure. When a storm coincides with a high tide and hits a low-lying coast, the storm surge can briefly lift the sea level like a giant high tide, and flooding may extend a long way inland. Widespread flooding in the Great Cyclone of February 1936, for example, was not caused by rivers bursting their banks, but by a storm surge. A house fell into the sea at Te Kaha, coastal roads were washed away and the sea swamped houses 100 metres inland at Castlepoint after the surge broke through sand dunes.

A rising sea level increases the potential for this sort of damage, but also has less immediate impacts. It can accelerate coastal erosion and flood groundwater systems with seawater, spoiling the water for agricultural use. Low-lying coasts that are liable to occasional flooding at very high tides could flood at every tide, and estuaries could enlarge as the highest level of tidal influence moves back upstream. Changes in sediment supply from rivers and in wave direction can have a big impact. In coastal Canterbury, a reduction in southerly waves could reduce the movement of river sand up the Pegasus Bay coastline, and could lead to up to 50 metres of shoreline erosion near the Waipara River mouth and up to 80 metres near the Waimakariri River. In places where coastal structures such as causeways and dykes stop the coastline from moving, or protect reclaimed land, the rising sea may eat away at beaches and the land. NIWA is advising local authority planners to allow for a 20 cm rise in sea level by 2050, and 50 cm rise by 2100. However, sea-level rise and in particular the rate of melting of the Greenland and West Antarctic ice sheets are the areas of greatest uncertainty at the moment, so those numbers may have to be revised upwards.

If we put all these projections together, it's clear that our climate is to going to change in ways that may not seem dramatic when expressed in degrees centigrade, but that will have a big impact on everyone living in New Zealand. Small shifts of climate can mean big changes at ground level.

5

Impacts:
the good and the bad

So far we've looked at global warming as a purely climatic phenomenon – changes in the sort of weather we can expect in the future, together with the direct impacts on ice and sea level – but it's what that climate change will do to humankind and all the other living things on the planet that really make it worth trying to avoid. Climate-change impacts can be obvious, such as a disappearing glacier or ice shelf, but they are also often subtle.

Ecosystems, the complex web of interactions between all the living things in a given place, are adapted to cope with the climate they're used to. Push them out of their "comfort zone" and things start to go wrong. Different species respond to climatic factors in different ways, affecting the overall balance of the system. For natural ecosystems, the rate of change is crucial. When the Earth was emerging from the last ice age, temperatures rose by about 5°C in 5000 years or so – a hundredth of a degree per decade. As the ice retreated, forests moved on to the newly exposed land, and animals followed. But today the climate is warming at 0.2°C per decade – 20 times as fast – and nature is having trouble keeping up. Ecological adaptation takes more time than the current pace of global warming allows.

To get an idea of what the plant life might have been like during the last ice age in New Zealand, you only have to look at the mountains. As you climb higher, the vegetation changes. Plants adapted to the coldest conditions eke out an existence near the permanent snowline. As the climate warms, that snowline moves further up the mountain. Plants suited to warmer conditions begin to invade from below, and the cold-adapted plants have to extend their range upwards. Eventually, if warming is strong enough, there will no longer be anywhere suitable for the cold-adapted plants to grow. Their ecological niche will disappear, and so will they, swamped by new plants better able to cope. There are signs that this is already happening – the treelines in Mt Aspiring National Park are rising. In Tongariro National Park, the introduced weed heather is beginning to spread higher than 1200 metres, formerly its upper limit. Decreases in the depth and duration of snow cover can also cause problems for animals that hibernate through winter under that snow.

Fig. 12 Vegetation zones on Mt Taranaki. As climate warms, the snowline will rise and warmer zones will spread up the mountain, displacing cold-adapted species. *Craig Potton*

Closer to sea level, climate change will enable warmth-loving species of plants and animals to live nearer to the poles. One recent estimate suggests that many species' ranges have been expanding towards the poles at about 6 km per decade for the last 40 years. On one level, this could be benign: if tropical fruits can be grown in Northland, few will complain, and olive groves might become quite popular in Dunedin or Invercargill. But if weeds and pests spread south, as some subtropical grasses are already doing in the North Island, there's more to worry about and potential costs to face. But other species can't move in order to stay in their preferred climatic zone, and national parks and wildlife reserves can't move south. In some places it may be possible to create climate corridors that will enable species to extend their ranges southward to match climate change, but towns, cities and agricultural

Fig. 13 Tuatara are threatened by climate change. Extra warmth during the breeding season can increase the proportion of males born, and some populations are already showing an imbalance. *Mike Aviss/Department of Conservation*

land act as barriers to free movement. Ultimately, species may face the same fate as the mountainside plants. Unable to keep up with change, or to get past physical or environmental barriers, they'll run out of room and be swamped by better-adapted species.

Extra warmth is also a threat to some iconic New Zealand species. The sex of tuatara when they hatch depends strongly on the incubation temperature of the egg. At 21°C, tuatara young have an equal chance of being male or female, but at 22°C 80 percent will be male. At 20°C 80 percent will be female, and if the incubation temperature drops to 18°C all offspring will be female. Tuatara are obviously better able to withstand a cooling of the climate (which they must have survived during the last ice age), because an excess of females in the population simply means that the males have to be busier breeders. But if there were a significant excess of males in the population, there would be fewer females to breed and the population would diminish. Some of the Cook Strait island populations of tuatara are already showing signs of a gender imbalance tipped towards males.

The royal albatross colony at Taiaroa Head, near Dunedin, is already feeling the impact of warming. The soil at the nesting site is thin, and lies on top of rock. In summer, the soil temperature can rise to 50°C, uncomfortably hot for the birds, so they stand up to keep cool. This exposes their chicks to flystrike and their eggs to overheating. The simple but ingenious response from the managers at the site has been to install a sprinkler system to cool the birds.

Temperature increases can also have less direct effects. Shorter winters and longer summers mean that spring comes earlier and autumn lasts longer. In Europe spring has been arriving earlier each year. Spain has seen early spring temperatures increase by 1°C per decade, and spring now starts two weeks earlier than in the 1970s. This could disrupt the relationships between species if they react differently to seasonal changes. For its first food in early spring, a bird may depend on a type of insect that feeds on a particular species of plant. If that plant responds to a shorter winter by growing earlier, it may get eaten by different animals, which means less food for the insect that the bird feeds on and therefore fewer insects for the bird to eat. If as a result the birds of that species on average lay fewer eggs, their population will decline.

Some animals have an inbuilt mechanism to judge the time of year by changes in day length. If such an animal feeds on another species that reacts to temperature signals, then the two can get out of sync. The one that's locked into day length will wake up and look for food that may already have been eaten by something else. Similar things can happen to migratory birds: by the time they've flown in from their winter feeding places, other birds may have mopped up the spring flush of food. Even migration itself can be disrupted by changing seasonal patterns. If autumn lingers longer, birds that would usually head off to warmer climes may hang around. When winter finally arrives they may be unable to cope, or their presence may increase competition for food and affect other species which stay year-round. Through all these processes, climate change affects the relative abundance of species, leading to a cascade of problems.

New Zealand's native beech forests occasionally have a really good year in which they produce huge quantities of seed. This means far more food is available than usual, and enables invasive species such as rats and mice to breed prodigiously. But the rodents don't just eat beech seed. They vary their diet with all sorts of things, including endangered birds such as the mohua or yellowhead, and when the beech seed runs out they will spread out, looking for food. There's good evidence that warmer temperatures are making this happen more often, and the Department of Conservation is concerned that it may need to increase its predator-control work to prevent more damage.

The species that do best in times of rapid change are those that are most adaptable. Plants that grow well and disperse their seeds over long distances can thrive when others fail to cope with the change. As species which are slower to adapt die out or fail to keep up with the moving climate zones, the ecosystem becomes simpler. The biodiversity is greatly reduced. In places where conditions stay relatively stable, such as tropical rainforests or coral reefs, ecosystems can become incredibly complex as plants and animals evolve highly specific relationships. The web of relationships can be vast, and trying to describe the ways in which the species all interact is next to impossible. The bits of the ecosystem we see – the larger animals and plants – are only the tip of the iceberg.

Apart from providing food, fuel, fibre and so on, ecosystems also provide humans with a range of services that are often overlooked and therefore undervalued. These ecosystem services include pollinating crops, giving shade and shelter, keeping water clean, controlling erosion and floods, and providing a pleasant living environment. The spread of human settlement and agriculture has had huge impacts on these services. Felling trees to open up farmland exposes soils to erosion. Draining marshes removes their ability to absorb floodwaters. Since European settlement began in New Zealand, 85 percent of the country's wetlands have been drained or extensively modified. Many of the remaining wetlands are coastal, and will be lost as the sea level rises. Once these sorts of things are lost, restoring them, if it's even possible, takes time and money. Direct human action is the most obvious way that ecosystems suffer, but rapid climate change is going to make things worse. Ecosystems and the services they provide are the natural context within which we live. If they are damaged, then we will feel the effects in our way of life and in our economic well-being.

Agriculture is all about managing simplified ecosystems to provide the food we need. Farming and horticulture are the first activities to examine for the direct impacts of climate change on human activity, and they are a huge part of our economy. Global warming presents challenges (and opportunities) for farmers and growers, and the way that agribusiness responds to those challenges will have a huge impact – not only on the New Zealand economy but also on the country's total emissions of greenhouse-gases and the way that we are perceived by the world.

Two factors dominate the effects of climate change on New Zealand's farmers: warmth and water. Too much or too little of either can hit farm production hard. The risk of flooding is likely to increase significantly over the next century, especially in high-rainfall areas already prone to flooding. Bad floods can ruin pasture, drown stock and damage fences and buildings, but if you could offer farmers the choice between flooding and drought, most would take the floods. Drought causes large drops in farm output and therefore incomes, and puts huge stresses on the farming community. Flood damage tends to be restricted to smaller areas, whereas droughts can affect whole regions. The loss of farm output during the severe drought of 1997–99 reduced gross domestic product (GDP) by 1.5 percent and cut more than a billion dollars from the national income.

Even under low-emissions scenarios the risk of drought increases significantly, and by the end of the century under higher-emission scenarios what we think of as a moderate drought might be an annual event in parts of the east coast of both islands. Over the next few decades we're unlikely to see dramatic evidence of that change, but in the second half of the century it will probably be inescapable. Drought could be the single biggest factor affecting agricultural output in New Zealand – unless we adapt our mixture of land uses to become less vulnerable.

Long, hot, dry spells also increase fire risk. We're already used to summer fire bans in many parts of the country, and roadside signs telling us about the current fire risk are a familiar sight. With hotter, drier summers in the east of the country, the potential for damaging fires will increase. The risk is greatest in forests, but fires in scrub, farmland and shelter belts have the potential to cause a great deal of damage and human misery.

In drought-prone areas the pressure on water resources is obviously going to increase. Water availability is already a contentious issue in many parts of the country. In Canterbury, a reduction in rainfall and increasing temperatures will probably be accompanied by an increase in the flows of the big rivers that rise on the main divide. The same weather patterns that bring the heat and drought to the Canterbury plains and foothills will also bring more rain to the West Coast and the main divide. Rivers which flow all the way from the main divide, such as the Rangitata, Rakaia, Waimakariri and Hurunui, could have good flows while the surrounding plains are in drought. Schemes to harvest and store river water have been proposed, but are highly controversial. Nobody wants to live under a dam or see good farmland flooded, but this is an issue that is going to be intensified by climate change. The politics of water will be a defining part of the rural scene over the next decades.

In one respect, continuing increases in the amount of CO_2 in the atmosphere may bring benefits. Plants use CO_2, and some plants will get a boost from the extra carbon in the atmosphere through a process called CO_2 fertilisation. This free aerial fertiliser may provide a small boost to agricultural productivity, but it's not a universal panacea. Not all plants respond the same way, and this variation in response can create imbalances that disrupt ecosystems. The downside of climate change caused by the extra CO_2 is likely to far outweigh the probable early benefit from increased agricultural yields.

Dairying is currently New Zealand's biggest agricultural business. The national dairy herd comprises about 3.5 million cows producing 13.9 billion litres of milk – equivalent to one billion kilograms of milkfat – 95 percent of which is exported. The dairy co-operative Fonterra is New Zealand's biggest company and the country's largest exporter, shipping 20 percent of total exports and accounting for 3.2 percent of GDP. The dairy industry faces a number of major challenges from global warming

Fig. 14 Dairying has expanded rapidly over the last decade, often replacing dryland farming and relying on heavy irrigation. *Neil Macbeth/The Press*

– including managing its emissions of methane and nitrous oxide.

Dairy cows turn grass into milk, so a major priority for any dairy farmer is to manage pasture to maximise grass growth. Over the next few decades, higher CO_2 levels and warmer temperatures could boost pasture yields by as much as 20 percent in some areas. This effect will not be evenly spread around the country. Areas that are currently cooler and wetter, such as Southland, will benefit more than those that are already warm and dry. Farms in traditional dairying regions such as Taranaki could benefit from a longer growing season and more moisture, though if there is a significant increase in heavy rain events this could cause losses through flooding.

Increasing CO_2 and warmth could also change the mixture of plant species in pasture. Some weeds and legumes will respond faster than the desirable grass species, perhaps reducing their yield. Subtropical grasses such as paspalum and kikuyu may spread south in the North Island. These are less nutritious than pasture grasses, but better able to withstand drought. Pasture management is therefore likely to become more demanding, even if productivity does rise at first.

Dairy farms in regions prone to summer drought, such as Canterbury, already rely on irrigation to maintain summer grass growth. Drier summers will put more pressure on irrigation-water supplies, even in years when there is not a drought. Longer irrigation seasons will also mean increased costs. Whether this will reduce farm incomes enough to encourage a change of land use is impossible to predict, but farms in regions with good rainfall and a lower risk of drought will probably continue to be the optimum low-cost dairy producers, and their economic advantage may increase.

Pasture growth is obviously also important for farmers producing sheep and beef cattle, and they will face many of the same challenges. Although the use of irrigation to boost production has recently increased in this sector, most farms rely on natural rainfall and are therefore vulnerable to any increases in the frequency of drought. Non-irrigated farms on the east coast of both islands are likely to be most affected, especially those on terrain that dries out readily, as in the hills south of Marlborough or in Hawke's Bay.

There are various means of coping with drought: early de-stocking to reduce pressures on feed, moving feed in from other regions, ensuring adequate stock water supplies, installing irrigation where possible, and so on, but if average rainfall declines enough to put farms into drought regularly (perhaps even annually) then farmers won't be able to hold out until the rain returns. This scenario is unlikely to be a major concern over the next few decades, but could become a serious issue in the second half of the century, when land-use changes may be the only option on some farms.

Arable and cropping farmers will face the same water-use issues but they will probably also get some benefit from warmer temperatures and increased CO_2, depending on the crops they grow. Maize, for instance, doesn't get much of a boost from more CO_2 because it metabolises carbon in a different way from most cooler-climate crops, but wheat and barley yields could improve by 10–15 percent by the middle of the century. However, these figures assume a continued plentiful supply of fertiliser and water. In practice, as cropping regions tend to have drier summers, the need to irrigate will increase – with all the problems that may bring.

Warmer climates will expand the range of some crops. Maize could become a significant crop in the South Island over the next few decades, and new crops such as rice, soybeans and sorghum could have potential in the north. Vegetable growers may also find that the range of crops they can grow in their region will change, and there could be significant benefits for crops prone to frost damage or that benefit from summer heat.

Tree and vine crops have a different range of responses to a warming climate, though water remains a crucial factor. Kiwifruit, for instance, require a certain amount of winter chilling in order to stimulate bud-break and flowering. This is also true for apricots, which is why they are grown in Central Otago and not Northland. Kiwifruit production around Kerikeri has declined significantly since the early 1990s, and one factor may have been more frequent warm winters. In the Bay of Plenty, currently the centre of kiwifruit production in New Zealand, some climate projections suggest that by the second half of this century the western subregion could have a climate as warm as that of present-day Kerikeri. Growers currently use sprays and other management techniques to minimise the impact of winter warmth on flowering, but under high-warming projections even this may not be sufficient, and production of the main kiwifruit cultivar, Hayward, could become uneconomic there after about 2050. The other major centres of kiwifruit production in Hawke's Bay and Nelson will not face the same problem, and could gain an additional benefit from reductions in the frequency of damaging spring frosts; but in both areas water availability could restrict expansion and reduce yields. If kiwifruit production is forced out of the Bay of Plenty perhaps other warm-climate tree crops could be grown there. There has already been a considerable expansion of avocado growing in the region, and citrus crops could also be more widely planted.

For the vineyards along the east coast of both islands, drier, warmer weather with fewer frosts to damage flowers and longer autumns to ripen the grapes sound like a perfect recipe for good wine, provided that there's enough water to keep things growing. A recent study estimated that the amount of land suitable for grapes could double over the next century. Vineyards could move south, right down into Southland. A major recent topoclimate study in Southland suggested that one valley might already be suitable for grape-growing. Warming will also mean that higher slopes that are currently too cold or regions that are too frost-prone will be able to grow grapes. Northern vineyards may have to change to grapes that prefer warmer climates, and the extra warmth may mean alterations in the styles of wine being made as winemakers react to changes in the flavour profiles and ripeness of the grapes they harvest. However, this relatively rosy outlook depends on warming being gradual and not too excessive. Too much warmth isn't good for making the high-quality wines that are New Zealand's niche in the business. One of the biggest trends in the Australian wine business over the last decade has been the establishment of new vineyards in cooler regions that can grow high-quality grapes, and therefore yield high-value wines, instead of the bulk, low-value wines produced in many traditional Australian winegrowing regions. The Aussies are finally discovering that there are limits to the market for alcoholic grape jam.

We've already seen how a warming climate can affect productivity and species mix in pasture, and cause weeds to spread. The same is obviously true for insects and other pests. As climate zones shift, species will follow. Pests once restricted to the north will become established further south. Some will respond to the warmth by producing more generations per season, meaning more crop damage. Willow sawflies, for example, take 108 days from egg to adult when temperatures are around 11°C, but do the same thing in only 22 days when temperatures are over 23°C. In Gisborne, sawflies can produce 5–7 generations in a year, compared with only 2–3 in Invercargill. Species that arrive from overseas may find a warmer, more congenial climate. Eliminating painted apple moth from Auckland was expensive and controversial, and similar biosecurity threats can be expected in the future. New Zealand is already home to 90 percent of the pest insects found in southeastern Australia, and nearly 80 percent of those found in similar climates in western Europe. There's a reservoir of pest species that could end up here, and with increasing international trade those insects are constantly challenging our biosecurity. A warmer climate is also likely to increase the number of potential pests, and they could damage our native ecosystems, agriculture, forestry and human health.

New Zealand doesn't currently have any of the mosquitoes or other biting insects that transmit dangerous diseases such as malaria, dengue fever or Ross River virus, but several species have tried to become established in the recent past. The world trade in used tyres provides a means for mosquito larvae to travel from country to country, and when they find a suitable home they can quickly become established. Salt marsh mosquitoes are a vector for the Ross River virus, a nasty disease endemic to Australia, and they nearly became established in New Zealand in 1999. The spraying campaign to eradicate them from Muriwai and near Gisborne cost almost NZ$2 million. Had they become established and bitten someone already infected with Ross River virus (such as an Australian holidaymaker) the disease could have become prevalent here. The risk of dengue fever in New Zealand is smaller, as the mosquito vector prefers Queensland levels of heat and humidity, which are not expected in this century.

A warmer climate can have a more direct effect on human health. Heat waves can cause an increase in death rates. The extremely hot western European summer of 2003 is estimated to have caused 45,000 deaths more than would have been expected in a normal summer. That sort of summer is expected to become more or less average in Europe by the 2030s. New Zealand's maritime climate protects it from most extremes of heat and cold, but heat waves do happen and are likely to become more frequent. In Auckland and Christchurch, an average of 14 people aged over 65 years die annually from heat-related causes; that number is likely to double for a one-degree temperature increase and reach 88 deaths per year if temperatures rise by three degrees. At the same time, warmer winters may bring a reduction in death rates from cold stress, but where the balance between summer and winter mortality rates will lie is not known.

Overall, global warming's impacts on New Zealand are a mix of the good with the bad. Agriculture is likely to see some benefits, at least over the next 20 years, and good forward planning may enable farmers to avoid the worst effects. Water is likely to be the biggest issue – too much in some places, too little in others. If warming is at the lower end of the range of projections we looked at in Chapters 3 and 4, New Zealand will suffer fewer adverse impacts. The amount of damage and area affected will therefore be less, though both are very hard to quantify at the moment. Even if warming proceeds along some of the higher scenarios, and if we can approach the changes knowing what to expect, we can adapt and probably do well.

We will, however, remain vulnerable to the effects that climate change has on our trading partners. If the global economy suffers, we will too, however well we are adapting to warming in the South Pacific. We have to limit the future impacts of climate change by taking steps to reduce our emissions (and by persuading others to do the same), while also working to prepare our adaptive response. It's not a case of "adapt or mitigate": we have to do both at the same time.

6

Sinking or burning: our Pacific neighbours

Whatever happens to our climate, we will not be able to escape the indirect effects of the climate change suffered by our neighbours and overseas trading partners. If Australia is hit by frequent droughts, fires and damage to iconic tourist attractions such as the Great Barrier Reef, New Zealand will feel the knock-on effects. People from low-lying Pacific islands threatened by rising sea levels could become climate-change refugees and look to New Zealand as their new home. Events across the Tasman and around the Pacific will play a significant part in shaping the politics of New Zealand's response to climate change.

Since 1950, Australia has warmed by 0.7°C on average. It has experienced more heat waves and fewer frosts. Rain has increased in the northwest but declined in the south and east, and droughts have increased in intensity. These changes are already causing stresses on water supplies and agriculture, changing ecosystems and reducing seasonal snow cover. New Zealand's climate is moderated by the oceans around it, but Australia is a large, dry, hot continent that stretches from the tropics down to mid-latitudes. Away from the coasts, the ocean's effects on its climate are much less marked. Australia is expected to warm at about the global average rate, but more markedly in the centre and the north of the country, and slightly less around the coasts and in the south, particularly during winter. There will be more very hot days and fewer cold days. With more droughts and greater extremes of temperature come the risk of more bush fires. When rain falls it is expected to be more intense (even where total rainfall is less), and to make drought and flooding more likely.

The IPCC's Fourth Report identifies a number of hot spots where warming is likely to have a severe impact on Australian ecosystems and economic activity. At Kakadu National Park, in the Northern Territory, saltwater intrusion into freshwater wetlands will cause mangroves to proliferate and fundamentally change the ecosystem. In the wet tropics of Queensland warming is expected to cause the extinction of native animals in the uplands, while sea-level rise will increase the danger of damaging cyclonic storm surges and coastal flooding. Reduced water flows in the Murray Darling Basin – up to 25 percent down by 2050, and as much as 50 percent by 2100 – will reduce water availability for irrigation and supplies to cities and towns, and wetland ecosystems will struggle. By 2020, there's a 50/50 chance the Murray River could be too salty for use as drinking water or irrigation. As Western Australia dries and water shortages intensify, native species will find their ranges reduced, and farmers are likely to struggle.

One of Australia's great natural wonders, the 2000 km Great Barrier Reef off tropical Queensland, is also at risk from climate warming. If sea temperatures rise above a critical point coral colonies can die off in what are known as bleaching events. Sea-surface temperatures over the reef have increased steadily over the last 100 years, and bleaching events have become more frequent since the 1970s. The massive coral die-offs during the intense El Niño of 1997–8 and in the summer of 2001–2 were associated with record high sea temperatures. Projections for the region suggest that bleaching events could be an annual occurrence by the second half of this century. The damage is expected to be worst at the southern end of the reef, where the corals seem to be most sensitive. The reef is also vulnerable to sea-level rise. If the sea doesn't rise too rapidly, the corals should be able to grow fast enough to keep up, but a sudden surge in the sea level could leave them behind. At the same time, increased ocean acidity caused by continuing absorption of CO_2 could make it harder for corals to build their skeletons. Warming and acidification have severe implications for the ocean ecosystems of the region, and from an economic perspective tourism could be badly hit.

Under all emissions scenarios, snow lines in Australia's mountains will rise. Recent work by the Commonwealth Scientific and Industrial Research Organisation (CSIRO) suggests that snow could almost cease falling in Kosciuszko National Park by the middle of the century. Australia's mountains are not as high as ours, and so the mountain ecosystems are at much greater risk as temperatures rise. The unique plants and animals have nowhere else to go.

Agriculture is not as important to the Australian economy as it was 50 years ago, but 60 percent of the land area is still farmed and much of it is vulnerable to more frequent droughts and less rain. The severe drought and extreme heat of the summer of 2006–7 brought a great deal of hardship to farmers in New South Wales and Victoria, and bush fires burned through large areas of southeast Australia and Tasmania. Drought in these areas is strongly linked with El Niño events, but the intensity of the drought and the heat suggested to many Australians that they were seeing the fingerprints of global warming (see Fig 15, p. 102).

Drought, extreme heat and fire are all expected to increase in frequency over the next century. There is already huge pressure on water supplies for both agriculture and cities and towns throughout the continent, and this will only get worse with climate change. To begin with, at least, Australian farmers may experience some benefits as CO_2 enrichment increases yields from pasture and grain, but these are expected to be offset by the damaging effects of higher temperatures later in the century. Any increases will also be limited by the diminished supply of water and nutrients, and the effects of increasing salinity. Although the frequency of extreme droughts is projected to increase, in regions where average rainfall decreases moderate droughts too could become much more frequent and possibly make traditional crops uneconomic. In the north, damage from tropical storms could also increase. Queensland's banana and sugar-cane crops are particularly vulnerable. Storms are expected to intensify as sea-surface temperatures increase, but it's not yet clear how significant this effect will be, or if their tracks will change.

If the countryside suffers, the cities will not escape. Sydney's average daily maximum temperature could increase by 1.6°C by 2030 and by as much as 4.8°C by 2070, and there could be a 40 percent reduction in rainfall by 2040 according to the CSIRO. By the 2050s, deaths of people over 65 years caused by heat stress could be eight times higher than at present. Demand for air-conditioners is likely to be high.

It's not good news for the Australian ski industry, either. Ski operators are already investing heavily in snowmaking to compensate for reduced snowfalls, but that may just buy limited time as snowmaking works only when the temperature is a few degrees below freezing. One of the classic signs of global warming is that night temperature increases more than

Fig. 16 Children play as waves flood over a road during high tide on Funafuti, Tuvalu. Tides were over 3 meters above normal early in 2005, causing extensive flooding. *Gary Braasch*

daytime temperature, because the greenhouse-gases capture some of the heat that would otherwise escape into space, and this will reduce the number of hours when snow can be made. Warmer winters will mean shorter skiing seasons as winter starts later and spring arrives earlier. Increasing numbers of Aussie skiers may head over the Tasman for more reliable conditions in Queenstown and Wanaka – a trend that's already becoming established.

In the tropical Pacific, sea-level rise makes the headlines, but the potential for increased storm and rain damage is also significant. Over the last 100 years the ocean surface temperature and island air temperatures have increased by 0.6–1°C, with more hot days and warm nights and fewer cool days and cold nights. Temperatures are projected to increase over the next century in line with the global average, by 0.5–3.1°C depending on the level of emissions.

Since accurate measurements of the sea level began in the early 1990s, it's become clear that the average rate of increase varies around the Pacific. In Tuvalu it's been 4.9 mm per year, in Kiribati 5.8 mm, and Nauru 6.7 mm, while the Cook Islands have an average increase of only 0.8 mm per year. The reason for this variation is that island sea levels are affected by the state of the El Niño/Southern Oscillation (ENSO). During an El Niño, sea levels in the western Pacific generally fall as easterly winds at the equator are weaker and don't push so much water across from the other side of the Pacific. During La Niña events, the easterly winds are stronger, the winds pile water up in the western Pacific and the sea level is higher. But predicting El Niño and La Niña events many years ahead is impossible, so island nations can't make simple, clear and confident predictions about their future. They can't simply draw a graph to show that they will be under water in, say, 2070 if the current average rate of rise is extrapolated. A lot of El Niño events could delay the problem, while strong La Niñas could hasten it.

That aside, generally higher sea levels increase the risk of damaging storm surges. The IPCC expects the number of intense tropical cyclones to increase, although the total might decline slightly. Some low-lying atolls can be completely inundated by a tropical storm that coincides with a high tide, wrecking homes and gardens and washing property away. When Cyclone Heta hit Niue in 2004, it had a huge impact on agriculture, turning the country from a food exporter to a food importer. Niue will take years to recover. Even if there are no severe storms, rising sea levels will cause seawater to penetrate groundwater and create salination problems in the soil of low-lying islands. Salt incursion can damage crops and contaminate drinking water. By the middle of this century many small islands are likely to be unable to cope with prolonged dry spells. Storm damage and erosion also threaten coastal settlements and important infrastructure such as ports and airports. On many islands the main road runs along the coast and is vulnerable to cyclone damage.

These problems have caused Pacific Island nations to be among the most vociferous supporters of action to curb global warming. In 2000 Tuvalu joined the United Nations, an expensive undertaking for a small nation, purely to lobby for action on climate change. Its strategy, according to prime minister Maatia Toafa, is to "continue making noises in international forums", and Tuvalu has been one of the most vocal of the island nations in the Alliance of Small Island States, a group of 43 states and observers from the Pacific, Caribbean and Indian oceans formed in 1990. The Tuvalu government is discussing immigration policies with New Zealand and Australia, and Toafa has said that his government would consider buying land in New Zealand or Fiji to resettle the whole nation if the islands became uninhabitable.

7

Warming in the wider world

Some of the impacts of climate change will be felt around the world, such as a sudden surge in sea levels. But temperatures will not increase evenly around the world. The impacts will vary from continent to continent and country to country. Living in the rural South Island, we may not experience the same direct impact of climate change as Western Europe, but if we want to sell our farm produce in European markets, the way that their economies are affected by that change will be very relevant.

The first and most important thing to note is that the greater the temperature increase, the greater the damage – and the more difficult it will be to adapt to the new conditions. This provides a context for action to limit future climate change and mitigate the expected damage. As we saw in Chapter 3, the IPCC's best estimate for a relatively low-emissions scenario is a rise of 1.8°C above the 1980–99 average by 2090, increasing to 4°C with a high-emissions scenario. That might not seem like a very big increase, but it's enough to make a huge difference to the world at the end of the century.

The planet is already 0.7°C warmer than in pre-industrial times. We're seeing the effects. The Arctic is warming fast, with permafrost melting and large reductions in sea ice; sea levels are rising 3 mm a year; mountain glaciers are melting and rock avalanches increasing; spring flows are starting earlier in rivers fed by glaciers and snow, and lakes and rivers in many regions are warming. Animals and plants are responding to changes in seasons and shifting their ranges towards the poles, and extreme weather events such as the European heat wave of 2003 are becoming more common. Satellite observations show that the spring greening is happening earlier in many regions as summers get longer. Ocean acidification will continue. Over the next 20–30 years we are going to see a further 0.6°C increase in average temperature. This warming is already "locked in" – it's a consequence of the way the climate system works and the amount of greenhouse-gas currently in the atmosphere.

The IPCC's latest report suggests that by the middle of the century water resources will increase by up to 40 percent in high latitudes and some wet tropical areas, but will decrease in other areas – many of them already desert or prone to drought – by up to 30 percent. Water supplies that depend on glaciers or snow cover will decline, reducing water availability for more than a billion people. In areas where water remains plentiful, crop production may get a boost from CO_2 fertilisation, provided that temperatures don't increase by more than 3°C. However, at lower latitudes, and especially in regions already prone to drought, crops will decline with even small local increases (1–2°C) in temperature. Carbon-dioxide fertilisation can boost yields only if crops have enough water and nutrients to keep growing. Extreme weather events such as very hot days will damage crops even if the overall climate remains favourable. In other words, farmers in the developed world may be able to benefit from CO_2 enrichment, while those in the developing world may struggle. Global food production should increase for a while, but if temperatures increase by more than 3°C, agricultural production will suffer everywhere. Fishery yields are also projected to decline owing to changes in the distribution and reproduction of fish as the ocean warms.

The great river deltas of Asia are likely to experience more frequent flooding and increased storm damage towards the end of the century as sea levels rise. These are some of the most densely populated areas in the world, and many are underdeveloped. People are poor and lack the resources to adapt to change on this scale, so the impact is likely to be severe. If the sea level rise is at the top end of projections, there is the possibility that hundreds of millions of climate refugees will flee their low-lying homelands. Poorer countries are also at greater risk of health problems because of global warming. On top of the direct impacts from drought and food shortages, flooding and storms, warming will change the distribution of disease-carrying insects, and pressure on water supplies is likely to increase the incidence of problems such as gastrointestinal infections.

Warming is going to have a large impact on natural ecosystems. Around a quarter of all the plant and animal species scientists have looked at are likely to be at increased risk of extinction if the global temperature increases by 1.5–2.5°C. There will be major changes in ecosystem structures and biodiversity is likely to be reduced. In a particularly worrying finding, the IPCC suggests that, after the mid-century, ecosystems will stop soaking up CO_2 and may start releasing it into the atmosphere, increasing warming.

Africa will suffer the worst effects of climate change. The area suitable for agriculture is expected to diminish. As early as 2020, up to 250 million people could be facing regular water shortages, and in some countries agricultural yields could be cut in half. Warming in the great African lakes will reduce the yield of the fisheries there. The potential for increased malnutrition and other social problems is clear.

Within 20–30 years, the melting of the Himalayan glaciers will reduce water supplies to huge areas of India, China and Southeast Asia. In the mountains, disappearing permafrost will destabilise slopes and cause rock avalanches. In parts of East and Southeast Asia, crop yields could increase, but in Central and South Asia they are expected to decline.

In Southern Europe, warming is projected to have a damaging impact on agriculture and tourism. Increased warmth will make Spanish and Greek beaches less attractive in high summer, and a reduction in rainfall will threaten water supplies that are already under pressure in some regions. In Central and Eastern Europe, summer rainfall will decrease, and summers will become markedly warmer. Forest productivity will decline.

In Latin America, the big threat is to the Amazon rainforest. By the middle of the century, warming and reduced soil water in Eastern Amazonia will lead to the replacement of tropical forest by savannah. As the region dries out, it could begin adding CO_2 to the atmosphere – a major feedback that will enhance warming in the second half of the century. In Peru, the melting of Andean glaciers is expected to reduce water supplies and cause severe agricultural and social problems.

North America will suffer water shortages that could reduce agricultural production. In the western mountain ranges, a decline in snow pack will reduce summer river flows and increase pressure on over-allocated water resources. Warmer, drier summers will increase the risk of forest fires and the area burned every year is expected to get significantly bigger. Changes in pest distribution will also affect forests. Heat waves are expected to become more common and more intense, and are likely to increase heat-related deaths among the elderly. The coasts of the Southern and Western US will be at risk from greater storm damage as a result of increasing sea levels and the potential for hurricanes to become more intense.

At as little as 1°C above the 1980–99 temperature, the Greenland ice sheet could pass the point of no return, where melting of the whole sheet becomes inevitable. The world would then face a sea level rise of up to 7 m over the next few hundred years – possibly much faster – with no means of avoiding it. If West Antarctica behaves in the same way, as much as another 5 m of sea level rise would be unavoidable.

At the top end of the IPCC's projections, agricultural production could become impossible over whole regions, increasing world hunger and potentially triggering a cascade of refugee movements. Australia would be particularly hard hit. The risk of abrupt, large-scale changes in the climate system would be greatly increased. At this point, humanity is far out into the unknown and the unknowable. With huge economic and social dislocation, will our civilisation be able to survive? I'd rather not have to find out.

This may sound a rather apocalyptic range of scenarios, but they are not, as the climate sceptics would say, alarmist. On the contrary, all these projections are directly from the IPPC's latest report, and are realistic assessments of what might happen under various climate scenarios. It is by no means a complete list, and it is not a set of predictions for the future. What it makes clear is that even moderate projections show that large habitats could be destroyed, the Arctic could lose sea ice in summer and the sea level could eventually rise by many metres. Refugees from climate-change-induced food and water shortages could cause or exacerbate regional conflicts. Specific events, such as a major drought or flood event, could trigger a cascade of social and political responses that would reshape our world. It's clear that, if we allow warming to get beyond a couple of degrees above the pre-industrial level, the world is going to be a very different place. The environmental, economic and social consequences are going to be enormous.

Can we avoid the worst of these effects? Over the next few decades we will see what a relatively small increase in the global average really means. If we can hit the lowest emission scenario envisaged by the IPCC, we will do well to be under a 2.4°C increase on pre-industrial temperatures by the end of the century – already well into the realms of severe damage. We also need to remember that this is a transient increase, not the equilibrium response. Warming will continue beyond the 2090s until the climate system gets back into balance. The challenge the world faces is two-fold: to adapt to the changes that are inevitable, and to come up with policies and technologies that will reduce our emissions enough to keep both the transient rise and the final equilibrium response below a level that will irreparably damage our civilisation. Adaptation and mitigation: both are essential.

8

Facing up to the inevitable

Global warming happens slowly from a human perspective, but for life on Earth it's happening at high speed. Climate change deals with averages and frequencies that change only over many years. I remember the 1960s, but apart from memories of a cold winter and some nice summer holidays I could not give an accurate description of the climate of the time, or how it compares with today. I could offer that old truism about the summers of childhood always being warmer, but as my childhood summers were often spent on islands off the west coast of Scotland, that certainly isn't true. One of my neighbours swears that the North Canterbury summers are colder than he recalls from his childhood and that the seasons have changed, but I'm not convinced. Human perceptions of change, unless accompanied by evidence, can't be trusted. We simply aren't good at remembering details of past seasons. Trees, on the other hand, do it rather well. Good growing seasons, with warmth and rain, can be seen in the width of their growth rings.

Our fallibility as climate monitors makes accepting the fact of climate change difficult for many people. When big changes are measured in tenths of a degree but daily temperatures can vary by 10 degrees, it's natural to feel that the small changes in averages can't be all that important. But as we've seen several times so far, little changes in some numbers can mean big changes elsewhere. The pace of change I've outlined means that in my lifetime – which if I'm lucky will extend into the 2030s – I'll see the evidence of dramatic change in the planet's climate system, but it'll be my children, born in the 1980s, and their children, yet to be born, who will feel the full impact of what preceding generations have done to the world. Global warming is an intergenerational issue, and that adds to the difficulty of deciding what, and how much, to do now.

The first stage of dealing with global warming is to face up to the inevitable change that's already built into the system. This is expected to be a change in the global average temperature of 0.6°C over the next 30 or so years, and the only thing we can do about it is to work out what effects it will have and devise ways of coping with them. This is adaptation, and it has to start now. But we also know that global warming isn't going to stop in 30 years' time, so our planning for the future has to include preparation for changes such as sea-level rise that are going to continue for centuries.

Adaptation is about coping with change. It has to include preparing for new events, and being resilient when they happen. Adaptation can occur on any scale, from local to national to global. On my truffière, I have already decided to improve my water supply and storage so that my truffles can cope with more frequent droughts. That will help me to avoid total loss of crop – or at least limit the size of the loss I'll face. My local authority might decide that the increased risk of heavy rain events justifies improvements to the stormwater system in the local town. Our regional council could encourage and coordinate management of water resources in our catchment, and national government might support work to identify land-use changes that could reduce vulnerability to diminished rainfall. At a global level, countries might work together to create new forecasting techniques that fill in the gap between weather forecasts and climate projections, that are designed to provide a long-range system that can give regional warnings of droughts or extended warm spells.

There is also a political dimension to the process of adaptation. Recognising that climate change is a real problem has proved a challenge for some of our politicians. On a regional level, some councils have been proactive and worked hard on defining their local responses, while others continue to regard climate change as a purely theoretical risk – one that might have to be dealt with some time in the future. Of all the challenges we face in adapting to climate change, getting political agreement on what to do is perhaps the greatest.

Adaptation also operates on different time horizons. A farmer might have a 10-year plan for his operation, while plant breeders developing improved crops will work on a 10–15-year planning cycle, foresters will be looking at 30–50-year timescales, and engineers building bridges or dams will look at 70–100-year lives (and longer) for their structures. The farmer can review his 10-year plan regularly, and respond to changes that he sees coming, but the designer of a dam is going to have to take into account *now* the changes that are expected to take place over the lifetime of the structure.

The adaptive capacity of human society is generally much greater than that of ecosystems. On one level, if a place becomes so prone to flooding that it is uninhabitable, the human population will move and rebuild. This might be an expensive and unpopular process, but it can be done. On another level, the social structures we provide for ourselves can make it easier to adapt to change. New Zealand's healthcare system is designed to provide medical treatment for everyone. If dengue fever were to arrive in Northland, dealing with it would not mean having to start completely from scratch. Health professionals already have contingency plans for dealing with outbreaks of new diseases, and a well-established network of health professionals to work on the problem. The medical response would also be matched by a biosecurity effort to get rid of the insects that carry the disease. Again, it might be expensive, and some people would suffer hardship and dislocation, but we could reasonably expect to be able to cope.

New Zealand is relatively rich and well equipped to adapt to many of the impacts of climate change. Poorer nations lack the systems, money and skills to adapt to global warming, and that is one of the main reasons why the IPCC projects that they will suffer more. Being able to cope with climate change isn't simply a matter of geographical good fortune, it also depends on how well organised you are.

Ecosystems can't plan for change, or work out what change might be coming. They can't be proactive and decide to move somewhere else. They can only react to changes, and that inevitably has impacts on the way they work and the services they provide for us. Adaptation to climate change therefore has to take into account more than the human perspective. We have to do what we can to limit the damage to ecosystems, and include measures to protect them from too much damage – if that's at all possible.

So in what ways is New Zealand most vulnerable? Increasing risk of drought in the east is certain to put water issues at the heart of any response to climate change. More intense rainfall in many regions will increase the risk of damaging floods. In the longer term, sea-level rise will affect coastal housing and industry, ports, beaches and ecosystems, and greater average warmth will bring significant change to where crops can be grown. How those risks and rewards are going to develop over the next 30 years is not very well understood at the moment. Developing that understanding is sure to be the focus of much research both here and overseas. The lack of detail in the medium-term "forecast" makes full cost-benefit analyses difficult. One severe drought in the next eight years would be bad, but not out of the ordinary. Two would be very unfortunate, and three disastrous for much of New Zealand's agricultural base, with knock-on effects throughout the economy.

The availability of water is going to be a critical issue, both for adapting to change and as a driver of local and national politics. Decreased rainfall and increased risk of drought in the east of the country will put pressure on water supplies for agricultural, industrial and domestic use. Town residents will not welcome being subjected to watering bans when out in the country massive centre-pivot irrigators are spraying water on to giant wheels of emerald-green grass. They will be even less pleased if run-off from those fields carries nitrates down into the groundwater that provides their household supplies, or if over-abstraction of that groundwater causes streams to dry up. Nor will they tolerate diverting water out of rivers, reducing flows and affecting wildlife and water sports. On the other hand, farmers will resent being told to switch off their water when they need it most. Losing grass growth will cost them money. Less milk sent to Fonterra will mean a drop in income for the farmer, a cut in exports and a reduction in the national income.

All these factors are currently being played out in Canterbury, where over the last 10 years there has been a rapid expansion of dairying on the plains. Summer grass growth there is unreliable at best, and evapotranspiration (ETP) rates can be very high. ETP is a measure of the rate at which plants extract water from the soil for growth plus the loss through their leaves, expressed as the amount of rain required to replace that water. With a hot nor'wester blowing, ETP can reach 12 mm

Fig 17 Maintaining summer grass growth in east coast areas depends on the extensive use of irrigation. *David Hallett/The Press*

per day at a time when rainfall might only average 50 mm per month. The soil rapidly goes into moisture deficit and plants stop growing. For the farmer who needs grass to keep cows productive, or has grain crops growing, water has to be applied early and often. Driving through the plains on a hot summer's day, you will notice that the amount of water being sprayed around is phenomenal. Large amounts evaporate before the water even hits the ground. Some of that water is sucked out of rivers and some comes from the groundwater under the plains – big aquifers fed by water from the foothills of the Southern Alps, which also supply Christchurch with the finest drinking water of any city in the world. Nitrogen fertilisers are used to promote grass growth, and some, usually in the form of nitrate, leaches down into the groundwater, from where it may eventually get drawn up again in domestic water supplies. In parts of Canterbury, the nitrate level in the rural water supply is already high enough to pose a health risk for pregnant women and babies.

Other farmers have begun to look for ways of further augmenting water supplies which are already over-exploited. A major proposal to harvest water from the Rakaia and Waimakariri rivers and store it for irrigation in a new man-made lake is highly controversial. The scheme

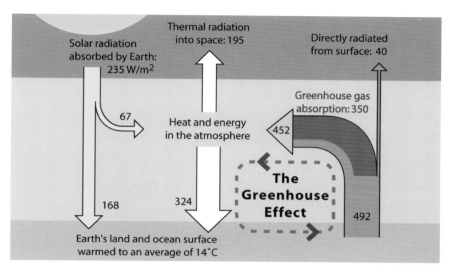

Fig. 1 How the greenhouse effect works. Radiation from the sun reaches the Earth's surface, heats it up, then is radiated back upwards at longer wavelengths and absorbed by the atmosphere, which in turn becomes heated. A lot of that heat is then radiated back downwards, adding substantially to the heating of the surface (see p. 16-17). *Global Warming Art*

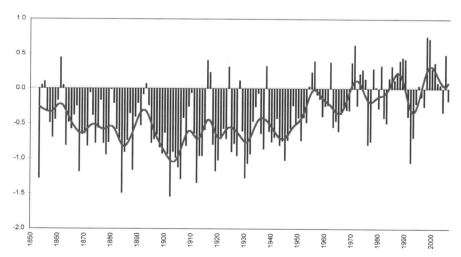

Fig. 2 New Zealand annual temperatures from the 1850s to date, shown as differences to the annual average for 1971–2000. The coolest years in the record occurred at the end of the 19th century. Note the significant swings from warm to cool years overlaid on a steady upwards trend (see p. 18). *NIWA*

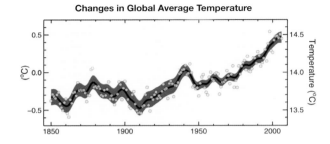

Fig. 3 Global temperatures from the 1850s show a steady increase since the 1970s (see p. 18). *Climate Change 2007: The Physical Science Basis, Summary for Policymakers, IPCC*

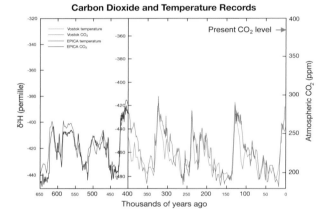

Fig. 4 Atmospheric CO_2 concentration and temperature over the last 400,000 years, as determined from Antarctic ice cores. CO_2 keeps pace with temperature throughout the last four ice-age cycles, during the warm interglacial periods. The present CO_2 concentration is more than 30 percent higher than any past interglacial peak (see p. 20). *Leland McInnes*

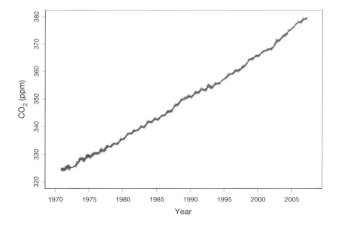

Fig. 5 Atmospheric CO_2 concentration recorded at Baring Head, near Wellington (see p. 24). *NIWA*

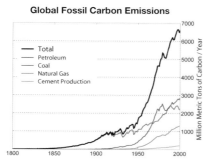

Fig. 7 Global carbon emissions from fossil fuels over the past 200 years. Until the early 20th century coal was the main fuel, but the rapid growth in use of oil and gas since World War 2 has boosted emissions enormously (see p. 34). *Global Warming Art*

Fig. 8 Sea-level rise since the last glacial cycle was at its cold peak. Note the rapid increase in sea level about 14,000 years ago (Meltwater Pulse 1A), and the relative stability over the last 6,000 years (see p. 44). *Global Warming Art*

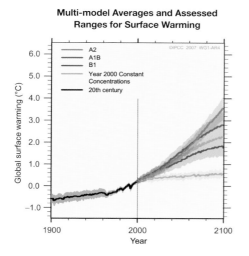

Fig. 9 Global temperature projections for the next 100 years for three key IPCC emissions scenarios. The A2 scenario (red line) suggests a world where regional differences remain strong and population increases to 15 billion by the end of the century. A2 and A1B (green line) are often called "business as usual" scenarios, because they are taken to represent likely emissions trajectories if the world takes no action on climate change. A2 is towards the top end of the range of emissions scenarios, whereas A1B is in the middle. B1's storyline suggests that population peaks at about 9 billion in mid-century and then declines, but supposes that the world economy shifts towards services and information technologies and that new efficient energy technologies are developed rapidly and massively used. Global solutions are sought for economic, social and environmental problems. B1 is at the bottom end of the IPCC emission range. The orange line shows the warming that would occur if atmospheric greenhouse gases had stabilised at year 2000 levels (see p. 49). *Climate Change 2007: The Physical Science Basis, Summary for Policymakers, IPCC*

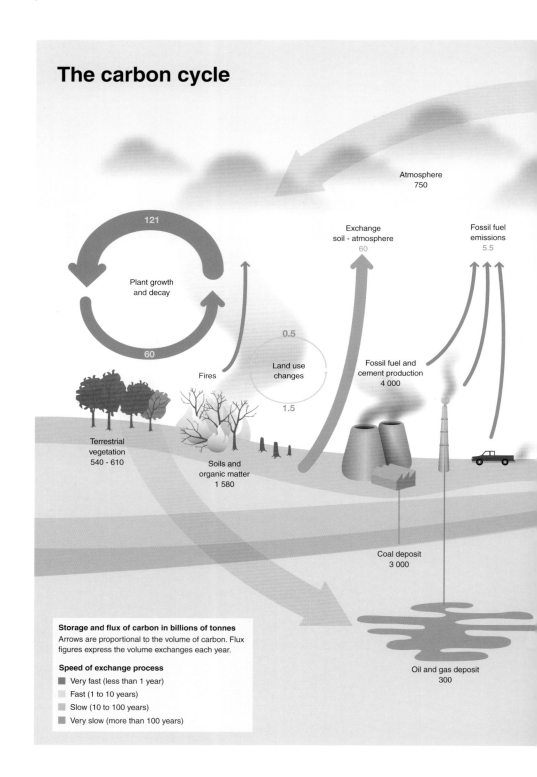

Global warming and the future of New Zealand | **101**

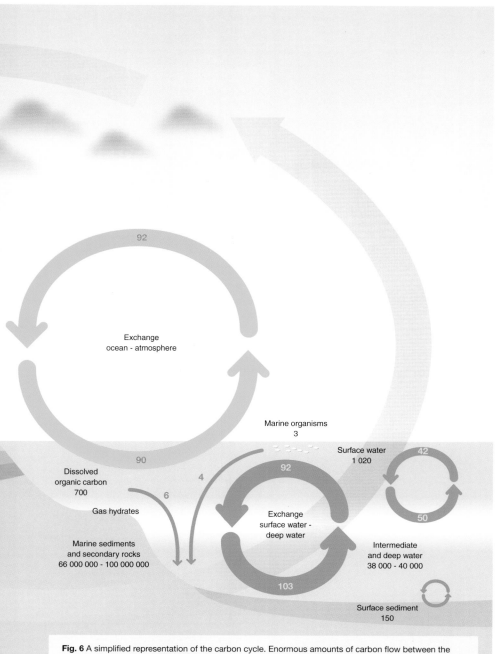

Fig. 6 A simplified representation of the carbon cycle. Enormous amounts of carbon flow between the oceans, the biosphere and the atmosphere, but before industrialisation this process was more or less in balance. Human effects including use of fossil fuels and land-use changes (especially deforestation) have caused carbon to accumulate in the atmosphere faster than it can be removed (see p. 33).

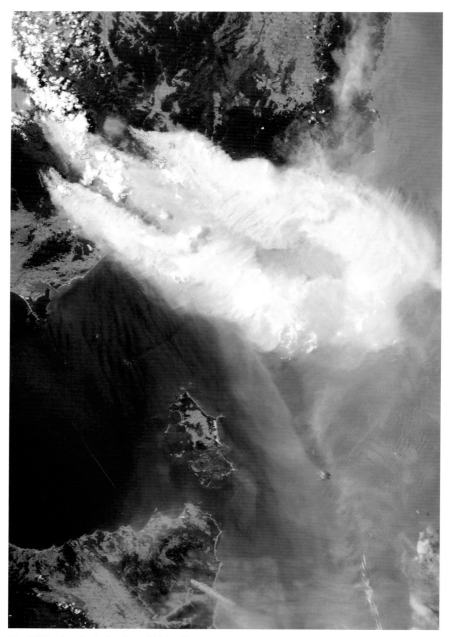

Fig. 15 Fires burning in the Great Dividing Range in Victoria send huge plumes of smoke out over the Bass Strait to the northeast of Tasmania in January 2007. Areas with active fires are outlined in red (see p. 80). *NASA*

Fig. 20 Global carbon emissions over the last 200 years, broken down by region. Europe led until 1900, when it was overtaken by the US and Canada. The oil shocks of the 1970s caused emissions to decline in both regions, but while emissions growth quickly resumed in North America, it flattened out in Europe. Note the recent rapid growth in China and India (see p. 124). *Global Warming Art*

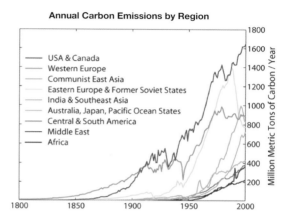

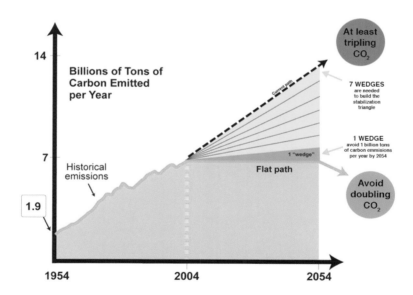

Fig. 21 Socolow & Pacala's wedges: the graph shows CO_2e emissions over the next 50 years. The top line shows a scenario where no steps are taken to constrain emissions. The bottom line represents stabilised emissions. The triangle between the two lines is what Socolow calls the stabilisation triangle. This is divided up into a number of wedges. Each wedge represents a single emissions-reduction technique, which starts out having a small effect, but as time goes by and its use expands (or technology develops) reduces more and more emissions. Each wedge represents a saving of 25 Gt C emissions over 50 years (see p. 126). *Socolow & Pacala*

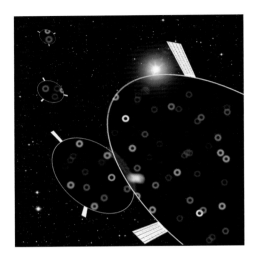

Fig. 23 Geoengineering as a climate solution: the solar shade proposed by Roger Angel at the University of Arizona would consist of trillions of little spacecraft flying a million miles above the planet, designed to reduce the energy reaching the earth by 2 percent – enough to compensate for a doubling of CO_2. The image shows how the little transparent craft would blur the sun's light (see p. 133). *Steward Observatory, UA*

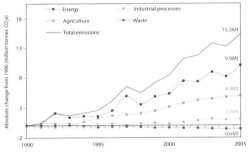

Fig. 24 New Zealand's CO_2 emissions have been growing steadily since 1990, with energy and agriculture accounting for the majority of the increase (see p. 136). *Government Climate Change Office*

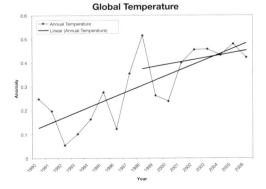

Fig. 29 Annual average global temperatures from 1990 to 2006, from the Hadley Centre. The trend over the full period is clearly upwards, even though 1998 is the hottest year in the period. Starting the trend line in 1998 (top line), makes no difference. It's still pointing upwards (see p. 175).

would harvest water from the big rivers over winter and spring when it is plentiful, and store it for when it is needed. This has obvious attractions – especially as global warming may boost flows in these rivers at the same time as it reduces rainfall on the plains. But such schemes create winners – the farmers who get the water, and losers – those who have their lands compulsorily acquired and flooded, or who have to live downstream of a large dam. Striking a balance between the two interests is a challenge at the best of times, and climate change will only make matters worse.

Irrigation on a large scale depends on plentiful cheap water, affordable power to pump it over the paddocks, and commodity prices high enough to cover running costs, repay the capital invested, and pay a good dividend to the owner. At the moment good money can be made, but are those farmers really meeting the full costs of their production? How vulnerable are they to changes in the cost and availability of water, or to measures taken to reduce nitrate leaching and run-off? Factor in the likely impacts of climate change, the risk of drought and reduced average rainfall, and the economics may change. And as we shall see when we look at how New Zealand can deal with its greenhouse-gas emissions, pastoral and dairy farmers have a role to play – and that might shift the economic balance even further.

The efficiency of water use also has to be taken into account. If the maximum sustainable amount is being used, the only way to improve productivity is by more efficient use of that water. Consider one hectare of land growing grass and another growing grapes. On a hot summer's day, the ETP might be 8 mm. To keep the grass growing the irrigation system has to deliver 8 mm of water over the full hectare. That's 80,000 L. Now consider the one-hectare vineyard with 3000 grape vines. On that hot day they might need 8 L of water each, delivered by drip irrigation. The vineyard therefore uses only 24,000 L of water – about a third of that used by the grass paddock – ignoring evaporation losses, which can be significant with spray irrigation. In terms of water use, not all crops are created equal. From the perspective of the regional economy and the community, is it better to have one dairy farm or vineyards over three times that area? This is obviously an oversimplification – dairy farms are not often in places where vineyards would thrive (though that might change as the Canterbury plains warm up) – but it does illustrate that the type of irrigation and crop can have a significant impact on how far a water supply can be made to stretch.

Hydroelectric power generation also requires large amounts of water, thermal power stations need water for cooling, recreational users – anglers, kayakers and white water rafters, boaties and swimmers – need good flows of clean water, and river ecosystems need adequate flows to maintain their health. Balancing all these uses is not a trivial undertaking. There have been legal disputes about access to water between agricultural users and power generators in the Waitaki catchment, and it is difficult to get planning consent for new hydro power schemes. As the country looks to increase power generation from renewable resources, pressure to expand generation on rivers with existing power schemes, and to build hydro schemes in new river systems, will increase. The health of our river ecosystems is not something we can take for granted, and damage there is not something we can ignore. New Zealand's trout fishery is one of the finest in the world and attracts many tourists. If the fishery suffers, so will the tourist industry and our image in the world.

All of these tensions are already there: the arguments and planning hearings and court cases are already happening. Climate change is simply going to add a new degree of urgency to the debate. In this case, adapting will involve setting up systems that encourage all the people with interests in water use in a given area to have a say in how the water is allocated and used – before major schemes reach the consent stage. If everyone who has a stake in the water and its uses is involved in the discussion about allocation, there will be a much better chance of arriving at equitable solutions. There is no simple answer, and the balance of competing arguments and the facts of the case vary greatly from river to river and region to region. We will need to be creative to find solutions, and we will have to build into the process an understanding of the changes that global warming will bring.

If droughts happen more often – and the projection is that they will – then farmers, farmers' organisations and local and national governments will need to have contingency plans in place. Each farm has to work out how to best survive a worst-case drought, while farmers can work together to create support services and organise supplies of feed. Local and national governments can help with money and staff time. Droughts can be slow to develop and patchy in distribution. A line of passing showers can ignore one property while delivering good rain across the valley. Detailed monitoring of rainfall and moisture deficits on a local scale can help with early warning of a drought, allowing farms to reduce stock levels and plan feed requirements. But prolonged dry spells are not always bad news. Different crops have different responses to drought. What's bad for a non-irrigated sheep farm can be great news for a vineyard. Dry conditions reduce the need for fungicide sprays, and increase the chance of a healthy, high-value crop. Eventually, if prolonged or regular drought causes pastoral farming to become uneconomic, changes of land use may be required.

Too little water is one problem, but too much is another. Climate projections suggest that the frequency of severe flooding caused by heavy rainfall is going to increase significantly over the next century, especially in areas where rainfall is already high. Furthermore, when rain does fall it will be more intense, even in areas where total rainfall is projected to decline, and this will increase the risk of erosion. Drought can worsen the run-off problem, because dry soils don't absorb water as well as moist ones. The worst rainfall events also damage all kinds of infrastructure, including buildings, fences, roads, bridges and railway lines.

Adapting to the increased risk of flooding and erosion involves assessing risks and planning to minimise them. Better early warning of heavy rain also helps landowners in flood-prone areas to move their stock to safety. Local and regional governments can assist landowners by providing maps of flood risk, and guidance about sensible responses to increases in that risk. An orchard that has flooded once might not cause the owner much concern, but regular inundation might make the operation uneconomic. In that case, the owner might consider switching

Fig 18 Heavy rain in Northland at the end of March 2007 brought 328 mm of rain to Kerikeri in 24 hours and led to dramatic flooding. *Jan Ravlich/Northern Advocate*

land use, or relocating. In Britain, damaging flooding in recent years has led to calls to restrict new building on flood plains, and the restoration of marshes and wetlands drained for farming. Marshes act as sponges, mopping up floodwater, smoothing the flow of water downstream, and are increasingly seen as an important part of flood control. Increases in insurance premiums in flood-prone areas can also provide an incentive for land and property owners to consider change and perhaps relocation.

Improvements to infrastructure can also help with increased flood risk. Upgrading existing flood-control channels, gates and dikes to cope with larger events is an obvious approach to managing bigger and more frequent flooding. But that sort of work can be expensive and take years to complete, and requires the people in the area to understand the need for the expenditure – to share the perception of risk. Rates increases are never popular. The best way to approach this problem is to ensure that as infrastructure is maintained it is also upgraded to take into account the impacts of climate change. Designers of a new subdivision can factor in the effects of heavier rainfall on the sewage and stormwater systems. Engineers and planners need to build the latest climate-change projections into the standards they use when designing urban, rural and transport infrastructure. If both size and return period of a very high river flow are going to increase, then bridges and other in-river structures such as power poles will need to take that into account. NIWA has already done a great deal of work in this area, and as climate-change projections improve these assessments will need to be regularly reviewed.

Over the next 30 years, warming in New Zealand is unlikely to cause major changes in areas where crops are grown. By the 2040s, however, the climate shifts will be large enough to have significant impacts. A good way to prepare for this is by conducting topoclimate studies – gathering detailed information about how climate varies with landscape. North-facing slopes are warmer than south-facing, higher slopes are cooler, hollows and valley bottoms can be frost-prone, and so on. By gathering weather data at key points throughout a region, comparing these with the local climate data, and then mapping the results with soil information and cross-referencing to a large database of crops, paddock-level information about crop suitability can be obtained. It provides an objective way of optimising land use – finding those crops that will provide the best return from any piece of land. If you take that topoclimate information, and then factor in the climate changes we expect over the next century, you have a means of seeing how the best crops to grow will alter with time. You can also adapt the parameters you use to judge the optimum crop to include water usage, for example. In a region with limited or highly contested water supplies, the optimum crop will be the one that produces the best economic return for the most efficient use of water. Detailed information about local climates allows rational decisions to be made about about crop choices.

Our present climate-change projections are probably not detailed enough to be of much help in this way, but there is plenty of scope for improvement and increased coverage of the country with topoclimate information – perhaps by encouraging (and funding) local and regional councils to commission the work. This a good example of an adaptive response that delivers immediate benefits – because optimising land use means increasing agricultural productivity, and therefore economic activity. It also provides the community with a tool it can use when looking at land-use change as a response to climate change. Changing what we grow can be more than an adaptation to warming – it can also be a way of reducing greenhouse-gas emissions, as we'll see in the next chapter.

Adapting to sea-level changes is another example of we are sure of the direction things are heading. NIWA suggested in 2004 that coastal planners should allow for a 20 cm rise by the middle of the century and a 50 cm rise by the end of the century. On the face of it, these numbers are in line with the base figures suggested by the IPCC, but they ignore the uncertainties associated with the melting of the great ice sheets of Greenland and West Antarctica. Over the next few years we will get a clearer picture of how the ice is responding to current temperatures, and how it might behave in future – and I suspect the estimates of sea-level rise will have to be revised upwards.

There are only three things you can do when the sea's rising: make a planned retreat, adapt or defend. The choice of response depends on a detailed understanding of local conditions, and how much money is invested in the current coastline and its uses. Planned retreat means allowing a vulnerable coastline to erode so that the natural shoreline moves inland. Buildings may need to be relocated, and planning controls used to prevent new construction in vulnerable places. Along most of New Zealand's coast, where farmland abuts ocean, this is the only practical option. Provided that the sea-level rise is not too fast, coastal ecosystems will change gradually rather than catastrophically with the rising waters. Shorelines will retreat, but there may still be a beach for holidaymakers to use. Wetlands and coastal marshes can spread inland – further up estuaries or across suitable flat land, There will be winners and losers – properties that lose land while others escape. Will the losers expect compensation? Leaving natural buffer zones between the sea

Fig 19 High tides and strong onshore winds combined to create high sea levels and large waves that caused extensive damage to houses and baches at Haumoana, Hawkes Bay in March 2005. *Tim Whittaker*

and new buildings and structures means that the risks from storm surge and flooding are reduced, if not eliminated. Managing coastal erosion requires the whole community to accept that the process is unavoidable, and that the changes are necessary even if painful and costly.

On coasts where there is already a lot of economic infrastructure such as offices, factories, port facilities and homes, adaptation to rising seas may be the best choice. This might include strengthening buildings against potential damage, or building floating structures in ports. Roads and railway lines can be raised above the expected flood level, the height of stop banks increased, and drainage systems improved. With care, land a considerable distance below sea level can be maintained as useful land, as Holland demonstrates. But even the Dutch recognise that it will be uneconomic to keep all their existing land dry. One response there has been to design floating houses – effectively multi-storey houseboats. Other adaptive responses may include actively creating wetlands and planting intertidal zones to act as buffers against storm damage.

Defending against sea-level rise is also part of the Dutch armoury, and may be economically feasible in a few parts of New Zealand

– particularly in high-value urban areas of Auckland and Wellington. Hard structures such as sea walls and breakwaters can be used to protect coastal infrastructure, but this usually means the end of beaches unless they are given regular top-ups with new sand. This has been done at Mission Bay in Auckland and in Napier, but is very expensive. New Zealand's main cities are all vulnerable to rising sea levels. Defence is likely to be the main response because of the value of the land and infrastructure, but it remains to be seen how it might be implemented.

If the impact of climate change in New Zealand is not all bad news, it's also important to recognise that adapting to change need not always be about costs and losses. A good example here would the response to the melting of the Tasman Glacier at Mt Cook. The new lake at the snout may signal the end for a large part of the glacier, but it has also provided tourism operators with a new opportunity. Boats now take visitors onto the lake to see the ice floes floating around and calving off the snout. Equally, continued expansion of the wine industry as warming opens up new regions to viticulture will bring significant increases in economic activity to those regions. And perhaps new world-beating wines for us to enjoy.

In all of the foregoing, it's obvious that early information about the coming changes helps people to plan suitable adaptive responses. GCMs are good at simulating the large-scale features of the climate system, and produce useful projections of how global climate will change as greenhouse-gas levels increase. They are not as good, however, at providing the regional detail that's so important to working out specific responses. Producing better regional information is a major priority for ongoing model development. As regional and GCMs improve, we will get a better picture of what lies ahead for the rest of the century, but there is still an important gap to be filled in the near term. Weather forecasting produces useful information out to 10–14 days maximum, and modelling of the El Niño / La Niña cycle can provide useful information about local seasonal trends. Weather forecasters and climate modellers are also looking at medium-range forecasting – producing seasonal and multi-year forecasts. Doing this successfully depends on a detailed understanding of the ocean – the distribution of sea-surface-temperature anomalies (which give us an idea of how much heat the ocean is moving around), and models of how they move (currents, eddies and gyres). What is in effect a "weather forecast" for the sea can then be used to develop a long-range weather and climate forecast. To do this, we need a lot of information about the sea to drive the computer models. Although the volume of satellite information has increased enormously over the last 10 years, modelling of the detailed circulation of the ocean system is still in its infancy.

The prospect of forecasting weather and climate on multi-year time-scales may still be remote, but when it is successful it will make planning for adaptation a much more reliable process. If a farmer can be warned that the next year will be dry, he can plan his crop or stock levels accordingly. If that forecast extends over multi-year periods, it may help to drive land-use changes. In this context, early adaptation can not only be cheaper, but it may reduce stress on landowners and allow communities to plan support strategies.

In the meantime, there is plenty that can be done that is both cheap and effective. A few years ago a study of how farmers on the east coast can adapt to climate change included detailed case studies and a series of workshops. Farmers from the Bay of Plenty, Gisborne, Hawke's Bay, Nelson, Marlborough and North and South Canterbury were presented with a scenario for mid-century that suggested increased warming, less rainfall and more risk of drought, and asked how they might best cope with that change. These are some of their conclusions:

> At the farm level, the focus should be on creating a more resilient operation that balances the economic activity on the farm with the ecology and the social well-being of the farm. Adopt a flexible approach that avoids putting too many eggs in one basket. Do what you can afford to do now, and keep the future in mind. Match the land use with the limitations and potential of the land. Celebrate successes. Plant trees for shelter and shade, for fodder and erosion control, and take degraded or less valuable land out of production and plant with trees. Improve water security through better storage and supply, and protect

against flood and erosion. Look to diversify farm operations through introducing new crops, or by boosting farm income through ecotourism initiatives. Manage soil well, and build fertility. Manage pasture with an eye to increasing flexibility, and balance that with a stocking policy that can react to change. Cropping farms should look to diversify their crop options and respond to changes. Manage the farm infrastructure to limit potential damage from flooding and erosion. Aim for energy efficiency and possible self-sufficiency. Focus on recycling and composting, and consider farm-based sewage and waste management systems.

That's a long list, but it's mainly commonsense land management. If farmers accept the reality of global warming and aim for sustainability, or adopt the old-fashioned (but highly valuable) policy of passing the farm on to the next generation in a better condition than when they inherited it, then much adaptation to climate change will take care of itself. If local, regional and national government can act as sensibly, global warming need not frighten us.

9

Cooling the future

Adapting to global warming is New Zealand's most immediate challenge. It's already too late to do anything to stop the warming that's built into the climate system, but if we do nothing to control the cause the world will continue to heat up. We will run the risk of severe and damaging climate change. Reducing carbon emissions from agriculture and burning fossil fuel will help to slow the future warming and limit the damage we'll do. This is called mitigation, limiting future damage, and means reducing CO_2, methane, nitrous oxide and halocarbon emissions. This is the stuff that makes the news. It's about carbon taxes and carbon trading, fart and fert (fertiliser) taxes, wind turbines, nuclear power, tidal generators, hybrid cars, public transport, carbon offsets, food miles, compact fluorescent light bulbs, zero-carbon houses, solar hot water and improved energy efficiency. It's driving international politics, as the world struggles to find a follow-on to the Kyoto Protocol.

The bad news is that there is nothing New Zealand can do that will make any difference to our own changing climate. Our carbon emissions are less than 0.5 percent of the global total, so even if we could achieve carbon neutrality, as the government would like, it would have no noticeable effect. We are wholly vulnerable to what the rest of the world does. It's therefore in the national interest to encourage international commitments to cut carbon emissions, and to try to ensure that those cuts are large enough to limit local damage from climate change.

Mitigation, like trade, is a global business. In 2005, total emissions of CO_2 from human activities were equivalent to 7.9 Gt C. The climate system mopped up about 2.7 Gt C of that, mostly through absorption into the oceans. This is all the Earth's present physical and biological systems can absorb, so if we wanted to stabilise the level of CO_2 in the atmosphere at 2005 levels, we'd have to cut our emissions by the difference between those two figures – which is 5.2 Gt C, or about two-thirds. Atmospheric CO_2 would stabilise at 380 ppm, and although the climate commitment means the Earth would continue warming for at least 30 years, we would have limited the total increase in temperature to about 1.6°C above pre-industrial levels. Human civilisation could then claim to be carbon-neutral, but the oceans would continue to become more acidic.

THE KYOTO PROTOCOL AND EMISSIONS REDUCTIONS

The world's first international treaty on climate change, the United Nations Framework Convention on Climate Change (UNFCCC), was adopted at the Earth Summit in Rio de Janeiro in 1992, and has been ratified by 154 countries. Its goal was to stabilise greenhouse-gas in the atmosphere "at a level that would prevent dangerous anthropogenic [human-caused] interference with the climate system". Two key questions were left unanswered: what constitutes "dangerous" interference and what level of greenhouse-gas in the atmosphere would cause it? It was a worthy objective, but the UNFCCC didn't have any teeth. Countries agreed to meet the goal, but there were no targets or mechanisms for achieving reductions. The Kyoto Protocol was an attempt to provide those targets and mechanisms. It was negotiated in Kyoto in December 1997 and came into force in February 2005. Developed countries that have ratified the Protocol are set emissions targets for the period 2008–2012 (the first commitment period) using their 1990 net emissions (the difference between the amount of carbon emitted in the country and the amount absorbed by carbon sinks) as a baseline. Developing countries could sign up, but would have no emissions targets. New Zealand's target is to keep emissions to 1990 levels. Overall, the Protocol was designed to achieve a global emissions reduction of five percent below 1990 levels during the commitment period. Emissions from shipping and aviation were excluded from the targets.

The key to the Kyoto Protocol is the mechanisms it provides to allow countries to meet their targets. These are designed to allow governments to choose between directly cutting emissions by, for example, replacing a coal-fired power station with wind farms, planting forests to create new carbon sinks, or by buying "credits" issued to countries that have undertaken emissions reductions schemes. Kyoto also created the Clean Development Mechanism, in which countries could earn carbon credits by investing in emissions reductions schemes in the developing world. Kyoto carbon credits are tradable, and so a government can decide whether it's cheaper to make investments in emissions reductions or to buy credits. This has created a global market in carbon credits, which the World Bank estimates to have been worth US$30 billion in 2006.

> Some US$5 billion was spent in developing countries under the Clean Development mechanism.
>
> At the time of writing, the USA and Australia were the only major countries that had not ratified the Protocol. India and China are part of the Kyoto process, but as developing countries have not been set targets. Negotiations for an international agreement to take over when Kyoto finishes in 2012 are underway. The key issues being debated are how to ensure that all countries – the USA, India and China included – take part in emissions reductions, and how to decide what targets should be set.

There's only one problem: reducing CO_2 emissions by two-thirds overnight would mean reducing global energy use by at least 50 percent. The lights would go out on a large part of the world economy. To describe this as an unpopular policy would be an understatement.

These are the hard realities of mitigating global warming. The higher we allow atmospheric greenhouse-gas levels to rise, the greater the harm the resulting warming is going to do to the planet and its inhabitants. On the other hand, reducing the amount of carbon we put into the atmosphere is not going to be easy. The global economy and the infrastructure of the developed world are built on burning fossil fuels. To change that is going to take a lot of work, time and money. There is also a very basic question that needs an answer: is it even possible to "de-carbonise" the global economy in time to stop enormous damage?

A pessimist would suggest it's already too late. We have given the climate system such a large jolt that it is committed to dramatic change, and nothing we can do will alter that. This is the position adopted by James Lovelock, the man who developed the Gaia hypothesis, which proposes that the biosphere itself regulates conditions on earth, keeping climate in the right range for life to thrive. Lovelock believes that in a 100 years or so humanity could be reduced to a few breeding pairs living in a warm Arctic. It's fair to say that this view is regarded as extreme. One kind of optimist might argue that climate change will be much more benign than we expect, so we don't need to do much. This is a position adopted by climate sceptics – or those with a vested interest in maintaining the economic status quo – and can't be entirely discounted. Based on what we know of the risks, however, it doesn't seem a very sound basis for global policy. As a strategy, it is like leaving your home uninsured: cheap and easy, but if you have misjudged the risk and the house burns down you lose everything. Another kind of optimist would say it's not too late: that if we work hard, implement new and improved technologies and have the political will, we can stabilise atmospheric carbon and avoid the worst effects of climate change. This optimism assumes the climate system is not going to unleash too many nasty surprises; in particular, that carbon-cycle feedbacks aren't going to make our efforts to cut emissions useless. It also assumes that we already have or can quickly develop the technology to reduce carbon emissions. And perhaps the most optimistic assumption is that the governments of the world can devise international policies that will achieve sensible emissions reductions. This cautious optimism is where the climate scientists, technologists, economists and politicians meet. It provides a scientific, technological and economic framework for action.

Within that position, however, there is room for degrees of optimism and different perceptions of risk. If you think the world is going to warm at the faster end of the range of projections, and that the impacts are going to be worse, then you will want more aggressive action to reduce emissions. If you believe that the costs of reducing emissions will be high, then you might argue for targets that allow higher emissions, meaning more gradual action. Deciding where you are along this continuum is not simply a matter of reading the science: it also depends on value judgements, ethical considerations and politics. This is where the real debate about global warming needs to be: in deciding the trade-offs between a credible response to climate change and the costs of, and methods used to achieve, the necessary reductions.

The aim of international action to mitigate climate change is therefore to arrive at a balance between what's feasible and what level of warming we are prepared to risk. We are already committed to hitting 1.6°C above pre-industrial levels. The impact of that is likely to be substantial. Most of the world's nations have not yet committed to any target, but the European Union (EU) has decided that it will treat 2°C above pre-industrial levels as the threshold for "dangerous anthropogenic interference" with the climate. (To be exact, they are aiming for 1.5°C above the average global temperature in the 1990s, which was 0.5°C above pre-industrial.) Because it's the only game in town, this has become the *de facto* standard. All the climate policies of the EU and its member states are intended to limit Europe's share of emissions by the amount required to keep temperature increases to that level.

I introduced the concept of CO_2 equivalent (CO_2e) earlier: it's the amount of CO_2 that would be required to have the same effect as the mix of CO_2, methane, nitrous oxide and halocarbons in the atmosphere. At the moment, it's about 455 ppm CO_2e, made up of 383 ppm of actual CO_2 plus other gases amounting to the equivalent of another 67 ppm CO_2. At the moment, industrial pollution and land use changes reduce the overall forcings due to greenhouse-gases to about 375 ppm CO_2e. To work out what level of greenhouse-gas reductions we should be aiming for, we need to run the global climate models at different greenhouse-gas levels to see what chance each level has of enabling us to get under our 2°C target. The key factor here is the climate sensitivity, as the amount the global average temperature is expected to rise with a doubling of atmospheric CO_2. The IPCC's best estimate is 3°C, but it ranges from 2–4.5°C. If it's at the low end of this range, we can get away with emitting more CO_2, but if it's high we have to be more aggressive in our reductions. At 450 ppm CO_2e (of which about 400 ppm would be CO_2), the chance of coming in under the 2°C threshold is about 50 percent, with a range of 75–25 percent. At 500 ppm CO_2e, the chance of hitting the target is worse, dropping to 30 percent, with a range of 5–50 percent. At 550 ppm CO_2e (roughly double the pre-industrial greenhouse-gas forcings, but with only 470 ppm from CO_2), the chances of achieving the target are worse still – under 20 percent. The best bet, 450 ppm CO_2e, is described in IPCC terminology as having a "medium likelihood". It's already too late to do better than that.

Let's look at those numbers in a different way. If we're going to try to stabilise greenhouse-gases at 450 ppm CO_2e, atmospheric CO_2 will need to be capped at about 400 ppm. At present we're at about 380 ppm, so we've only got 20 ppm to go. In recent years, CO_2 has been increasing by about 2 ppm per year, so if we do nothing we'll hit 400 ppm in 10 years or so. In terms of total carbon output, 20 ppm is 40.26 Gt C. If the planet is going to soak up about a third of our emissions, that means we can emit about 60 Gt C before we reach the danger point. If we allow greenhouse-gases to rise to 550 ppm CO_2e we will almost certainly miss our 2°C threshold (the climate sensitivity suggests we are much more likely to hit 3°C) but the emissions target will be easier to meet. Few people are seriously suggesting that atmospheric CO_2e be allowed to rise higher than 550 ppm. That prospect is too scary.

A lot of work is therefore being done to find emissions trajectories that are both feasible and affordable, and which minimise both the transient and equilibrium temperature rise. This is a highly complex field, beyond the scope of this book, but is a critical part of designing the framework for emissions reductions in any post-Kyoto treaty, and is covered in depth in the IPCC's Working Group Three report.

An important factor that should motivate action on emissions is what's been called the procrastination penalty. This means that the longer we wait to take action, the bigger the price we pay in both temperature rise and the size of the emissions cuts we will eventually have to make. We are already paying this penalty, as the IPCC makes clear in its 2007 report. If the world had taken concerted action during the 1990s and been able to limit greenhouse-gases at their 2000 levels, we would now be experiencing temperature rises of 0.1°C per decade and global temperature would peak in another 20 years. Instead, we are warming at 0.2°C per decade, with no end in sight. The procrastination penalty is the price we pay for inaction, and it keeps going up.

One of the standard arguments used by climate sceptics and other opponents of the Kyoto Protocol is that the cost of reducing emissions will be so great that it will cause huge damage to the economies of the developed world. They say Kyoto is going to achieve so little, and at so high a cost, that it's not worthwhile. The US and Australian governments took this line and are the only major countries outside the protocol's mechanism. Kyoto was established to enable countries to arrive at a least-cost solution to achieving some modest greenhouse-gas reduction targets, to establish a mechanism for larger emissions reductions in the future, and to involve as many countries as possible, including rapidly developing countries such as India and China. Each nation could either cut its emissions or buy carbon credits on an international market to meet its negotiated Kyoto target. In other words, if it were cheaper to buy carbon credits from a country that had made emissions reductions, then you could do so rather than implement reductions in your own country.

Kyoto was never intended to be the final answer. It was a start, and the international carbon market it created was explicitly designed to enable countries to reduce emissions cost-effectively. Whatever framework is put in place after the final Kyoto commitment period ends in 2012, it is likely to use a similar mechanism. The EU already has an active emissions-trading market which could form the nucleus of an international scheme, and the New Zealand Stock Exchange has announced that it will start its own Australasian carbon market in 2008. The idea that emitting carbon should have a price is designed to send a signal to the world's carbon emitters (the market), which will respond by seeking to reduce their exposure to this cost, so emissions will be cut. (It is perhaps ironic that an international system so explicitly designed to use market-based economics was rejected by the US, home of that economic theory.) For it to work, however, all emitters have to be exposed to the full cost of their emissions, and designing the rules and regulations to achieve that is not simple. Economists generally prefer a flat charge per tonne on CO_2 emissions – a carbon tax. This increases the cost of carbon-intensive goods and services, and provides a clear signal to producers and consumers alike: set the charge at the right level and the market does the rest.

Governments, however, generally prefer what are called "cap-and-trade" systems, because they provide a means to avoid penalising big emitters when the system is first introduced. Such systems are based around emissions allowances or credits that can be bought and sold. The cap is the total amount of emissions the government deems sensible. Companies are required to have credits that cover their emissions, and can either buy them in the market, or earn them by making emissions cuts. Businesses can decide therefore whether it's cheaper to buy credits or invest to reduce their emissions. If the cap is gradually reduced over time, the cost of emissions increases and the market should respond by making reductions. This is the basis of the Kyoto system, and similar systems have been used around the world to reduce pollution.

Governments can also cushion the impact of introducing cap-and-trade systems by giving big emitters free credits. A large coal-fired power station might become uneconomic if suddenly exposed to the full cost of its carbon emissions, but be essential to the national power supply. The government might therefore give the power station free credits to cushion the impact. This is called "grandfathering". However, it is fraught with problems. In the European carbon-trading system, for instance, big emitters handed free credits simply sold them for a quick windfall profit. Despite these issues, cap-and-trade systems are almost certain to be at the heart of national and international systems of emissions reductions.

One of the key issues in designing a global system to reduce emissions is to make it fair. Countries of the developed world have achieved their current standards of living by making use of cheap fossil fuels. Roughly three-quarters of the CO_2 that's been added to the atmosphere since pre-industrial times has come from the developed world. The US is responsible for about a quarter of all current emissions (though it will soon be overtaken by China). Unfortunately, the countries that are now growing fast, and which aspire to western-style standards of living (principally India and China) can't get there in the same way. If their rapidly expanding energy needs are met solely by fossil fuels, then the world simply cannot avoid damaging climate change. Nor is it morally or ethically defensible for the developed world to expect the billions of people in the developing world to give up their chance to achieve the lifestyle we already enjoy. The Kyoto Protocol explicitly recognises this.

It states that individual countries' commitments to emissions reductions should reflect their different levels of development and greenhouse-gas emissions patterns, as well as the principles of "equity and common but differentiated responsibilities". In other words, we need India and China on board, and if that means letting them emit more carbon while the developed world makes cuts, then that's what we'll have to do (see Fig 20, p. 103).

One approach that has been suggested is called "cap and converge". It entails deciding on a figure for the maximum amount of carbon emissions that can be sustained annually. This is then divided by the global population to give an amount that each person on the planet is allowed to emit. At the moment, the citizens of the developed world will be emitting far more than their personal allowances, and those in developing countries far less. Over time, the developed countries implement emissions reductions, while the developed world can increase their emissions. This seems inherently fair, and is likely to be popular in developing nations, but is not likely to go down well with the countries that have to make the biggest cuts. Exactly how the relationship plays out between the developing and developed world will be critical to implementing an emissions trajectory that produces the desired climate result. The negotiations (and the role the US plays in them) are going to be a key part of international politics over the next five years. The targets they set will decide our future.

The encouraging news is that the latest assessment of the cost of action suggests it will be much less than the cost of dealing with the damage caused by doing nothing. Economists may quibble about the details, but the Stern Review, a 2006 report by British economist Sir Nick Stern for the UK government and considered to be the most authoritative and complete examination of the economics of climate change, found that "the benefits of strong, early action on climate change outweigh the costs". And in a comment that was not lost on politicians of all persuasions, Stern suggested that "tackling climate change is the pro-growth strategy for the longer term, and it can be done in a way that does not cap the aspirations for growth of rich or poor countries. The earlier effective action is taken, the less costly it will be".

Stern looked in detail at the costs of stabilising greenhouse-gas levels at between 450 and 550 ppm CO_2e. These he put at roughly one percent of global economic activity. Stabilising at or below 550 ppm CO_2e requires that emissions peak in the next 10–20 years then fall by 1–3 percent per year. In 2050, emissions would have to be 25 percent below current levels. To stabilise at 450 ppm CO_2e, emissions would have to peak within the next 10 years, and then fall by more than 5 percent each year. In 2050, emissions would need to be 70 percent below current levels. Achieving the lower emissions target would enable the climate to stay within the EU's 2°C threshold, but would be more expensive.

Regarding the damage that would be done by not controlling emissions. Stern's review made world headlines. He found that "business-as-usual" emissions scenarios would cause climate change that would force a reduction in consumption of 5–20 percent per head of population. It sounds like the classic justification for taking out an insurance policy: pay a small price now, or risk losing a lot more later.

The IPCC's latest Working Group 3 report supports Stern's general conclusions about the costs of mitigation. It projects that meeting a target of between 435 and 535 ppm of CO_2e would only cost about 3 percent of global economic activity, or about 0.12 percent a year over the period to 2030.

Burning fossil fuels for energy is responsible for about two-thirds of all CO_2e emissions. Land use, agriculture and waste account for the other third. Fossil fuels produce mainly CO_2, but emissions from other sources also include a substantial amount of methane and nitrous oxide. Power generation is responsible for about a quarter of total emissions, and transport for about 14 percent. There are four basic ways to cut these emissions: reduce demand for goods and services that produce high emissions; increase efficiency of energy use; reduce emissions from other sources such as agriculture or deforestation; and adopt low-carbon technology for power, heat and transport.

We will have to use a wide range of methods to cut global emissions, and countries will use different combinations of methods depending on their circumstances. We'll look at New Zealand's position in the next chapter, but for now I want to focus on the global picture and introduce the concept of emissions "wedges". This was an idea developed by Princeton academics Robert Socolow and Stephen Pacala in the 1990s, and describes how a variety of emissions reduction techniques can combine to push emissions down to the point where greenhouse-gas levels stabilise, without dramatic impacts on economic growth (see Fig 21, p. 103). Their wedge concept demonstrates that if you use a range of different emissions-reduction methods, starting on a small scale but steadily increasing, it is possible to stabilise atmospheric CO_2e without having a severe impact on economic growth. A typical wedge might be to stop all deforestation, or replace 1400 large coal-fired power plants with gas-fired plants (because burning gas emits less CO_2 per unit of energy produced than coal), or make a 25 percent efficiency gain in the use of electricity in homes, offices and shops, or double today's installed nuclear power generation. There are many more possibilities, but the central idea that you need a wide range of actions, each starting small but implemented consistently over 50 years is a useful one to bear in mind as we run through the global options for action. The job is going to be difficult, but it can be done.

We can reduce emissions from power stations by using less electricity or by using less carbon-intensive means of producing it. The first and most obvious step is to use generation methods that emit little or no CO_2. Nuclear power is controversial in many countries, but will certainly play an increasing role in world power generation, because it is (after construction) effectively emissions-free. France generates around 80 percent of its electricity in nuclear power stations, and many countries are planning to build new nuclear plants, particularly the US. Some so-called renewable power sources have their drawbacks. Hydroelectric power generation is effectively emissions-free (except during construction), but depends on reliable rainfall to keep the storage lakes full. Geothermal power stations are reliable and cost-effective, but there are very few places where conditions are suitable to build them. Wind turbines can provide a considerable amount of power when the wind's blowing, but are considered unsightly or noisy by some. Solar cells and other solar technologies offer a virtually unlimited supply of power, as long as the sun's shining. Extracting power from tides and currents in the sea will eventually provide reliable clean power. Renewable energy sources are the fastest-developing sector of the global energy industry, and will become hugely important in reducing our carbon emissions. Improving these technologies and reducing the cost of the power they generate both will be an important focus of research and development over the coming decades.

In the short term, burning coal, oil and gas to generate electricity will need to continue, so various techniques to clean up their emissions are being developed. Switching from burning coal to gas produces more power per unit of emissions, but carbon still enters the atmosphere. Capturing and storing the CO_2 released when coal is burned – called carbon capture and storage or CCS and often described as "clean coal" – is feasible but has not yet been tried on a large scale. It depends on stripping CO_2 from the waste gases as coal is burned and then storing it so it cannot leak back into the atmosphere. Old oilfields are the most likely storage sites. Gas-fired power stations are much more efficient than present coal plants, especially when used in combined heat-and-power installations (where waste heat from the turbines is used to heat nearby buildings) but, like oil- and coal-fired power stations, they still emit carbon and are vulnerable to supply shortages and price volatility.

Technological innovation in power generation and distribution is going to be very important to any programme to reduce national or global carbon emissions. Some technologies could be close to breakthroughs that would radically change the way we generate energy. If photovoltaic cells (which make electricity from sunlight) can be improved in efficiency and reduced in cost, then it might become economic for every home to have a panel, taking much of the load off network power generators. This concept, called distributed generation, is already becoming popular. Homes remain connected to the national grid, but only draw on it when their own power sources such as solar panels or small-scale windmills can't provide enough energy. At other times, the small generators sell their excess power back to the grid, effectively running the power meter backwards and saving on power bills. If solar cells could be made cheap enough to use as wall or roof cladding, buildings in many parts of the world could become self-sufficient for energy. Bioengineers are working on systems that mimic the way that plants capture the energy from sunlight, a process much more efficient than the best solar cells, holding out the possibility of truly "green" power. None of this is science fiction: it's all being worked on in labs around the world. The sooner the research is translated into practical products, the better. But it doesn't mean we have a reason to wait for the next big thing. The procrastination penalty is already being paid, and we have to work with the technologies we've got.

Making better use of the energy we generate is an obvious first step. Designing homes that require little or no winter heating or summer cooling is not new. A German housing design approach developed in the 1980s, the *PassivHaus*, costs only a little more than a normal home to build but its power consumption is only about 20 percent of that of a standard home, and none of that goes on heating or cooling. In London, Britain's first carbon-neutral housing estate is being built. The complex will have windmills and solar panels to generate electricity and burn wood for heat and to generate extra electricity. Rainwater will be collected and used to flush toilets and water plants. Solar water heating can reduce household electricity bills by 40 percent. Compact fluorescent light bulbs use only a quarter the electricity of traditional incandescent bulbs and last 10 times longer. Efficiency improvements are an essential component of any strategy to reduce carbon emissions and the impact can be enormous.

Reducing carbon emissions from transport fuels is more difficult. Perhaps the most obvious method is to reduce the number of vehicle-miles being driven. Railways are much more fuel-efficient for freight and passenger transport, so policies to encourage their use and get many cars and trucks off the roads would be beneficial. Investment in public transport also makes sense. Liquid fossil fuels are very energy-dense (meaning they have a lot of energy per litre, so a little will take you a long way) and there are not many obvious substitutes. A few approaches can directly reduce carbon emissions. You can dilute the fossil content of the fuel by adding biofuels – alcohols and oils derived from the agricultural production of crops such as maize and sugar. These are carbon-neutral because the carbon in them was taken from the atmosphere, so when they're burnt there's no net increase in atmospheric carbon levels. The one major drawback is that growing crops to provide liquid fuels takes up a lot of land. In the USA, the rapid increase in use of maize as a feedstock to make ethanol has already forced up corn prices. Mexicans, for whom corn is a staple, have seen tortilla prices increase by 120 percent between 2005 and early 2007. On a global level, competition between crop production for biofuels and for food will put pressure on supplies of both.

Another approach is to make vehicles go further on the same amount of fuel. Petrol and diesel have been cheap and have no emissions cost built into their price, so there has been little incentive to improve vehicle efficiency. (In fact, average vehicle fuel efficiency in the US has declined since the 1970s.) However, good engine design coupled with reductions in vehicle size and weight can provide huge fuel savings. A German company, Loremo, is building a car which will use only 1.5 L per 100 km. So-called hybrid cars have an electric motor and batteries as well as an ordinary petrol engine, and can run on either; their fuel efficiency can be impressive, especially around town.

Aviation faces a major problem in reducing its carbon emissions because there is no feasible alternative to kerosene as jet fuel. Aircraft mostly discharge their emissions high in the troposphere where the combined warming effect of the exhaust gases is greater than just the CO_2 alone. Jet engines are increasing in efficiency, which has helped to improve the emissions per passenger-mile, but this has been more than offset by a tremendous expansion in air freight and passenger travel. Aviation emissions are not counted as part of any country's emissions, so they escape the Kyoto process, and there has been pressure, particularly in Europe, to bring them into an emissions-trading scheme. How this will impact on airfares and international travel in general is not yet clear, but it might go some way to explain why Sir Richard Branson, owner of Virgin Airways, has offered a US$25 million prize for a scheme that can remove large quantities of CO_2 from the atmosphere. The aviation industry is keen to establish its own system of carbon offsets and trading, perhaps calculating that this may help avoid more drastic action. Aviation is reckoned to account for about 2.5 percent of total emissions, but shipping, which has seen enormous growth in the last decade as world trade has boomed and is also not covered by the Kyoto Protocol, is estimated to account for a further 5 percent. It is also likely to face pressure to reduce emissions.

In the long run, it may be possible to wean much of the world's transport system off the use of carbon-based fuels, leaving biofuels to handle the areas where substitution is impossible. Much work is being done on using hydrogen as a transport fuel, either in lieu of petrol in an internal-combustion engine, or as fuel for a fuel cell, a device which generates electricity by combining hydrogen with oxygen from the air. Unfortunately, hydrogen is a lot less energy-dense than fossil fuels, and poses major problems of storage, handling and distribution (it is best kept as a liquid at minus 253°C, which means a lot of power is needed just to keep it refrigerated). It is not always emissions-free: most commercial hydrogen is made by processing methane from fossil sources. However, it can be made by passing electricity through water, a process which is 100 percent emissions free, as long as the electricity comes from carbon-neutral sources such as sunlight.

Using electricity in transport is hardly new. Electric cars were as common as internal combustion engine cars a century ago. Electric motors can be very powerful, and the braking system of an electric vehicle can regenerate electricity, which is fed back into the batteries, helping to make it very efficient. Unfortunately, the batteries available are usually heavy, have limited power storage, and take a long time to recharge. Battery technology is moving fast though, and some very exciting electric cars are entering production. The Tesla Roadster, an electric sports car designed by Lotus in the UK for Tesla Motors in California, offers a 400 km range using lithium batteries (the same sort

Fig. 22 The Tesla Roadster: US$100,000 buys supercar performance in an electric vehicle. *Tesla Motors*

used in laptop computers) and will accelerate to 100 kph in around 4 seconds. Provided that you don't need to do more than 400 km a day and buy your electricity from a renewable source, you have emissions-free transport – for US$100,000.

Other types of batteries may offer very short recharge times. Ultracapacitors are devices than can charge (and discharge) very quickly, and a number of companies are working on using banks of these for use in cars. Flow batteries use a chemically charged liquid as a fuel. As the battery discharges, spent liquid is replaced with fresh. This technology is being used to store electricity generated by wind farms, and gets around the problem that wind farms can only provide power when the wind's blowing. A flow battery charges the liquid in it and pumps it out to a holding tank while the wind is blowing. If there's no wind when the power is needed, the flow battery discharges, pumping the fuel back out of storage.

Agricultural emissions of greenhouse-gases amount to about 14 percent of the global total, so they can't be ignored when planning emissions-reduction strategies. Methane is emitted by the animals used in pastoral farming (usually as burps, not farts) and also by rice farming. Nitrous oxide is produced by the decomposition of animal wastes and fertiliser, and CO_2 is emitted from soils, burning stubble and other farm activities. Conservation tillage (sometimes called no-plough farming) significantly cuts carbon emissions and would save 25 Gt C over 50 years if all the world's cropping farms adopted this practice – enough to qualify as one of Socolow and Pacala's stabilisation wedges (see fig 21, p. 103). Changes in rice-growing methods can significantly cut methane output, particularly use of cultivars of rice that can be grown in dry (or drier) conditions. Nitrous oxide emissions can be reduced by using just enough fertiliser to get the required growth, so there is no excess to run off and decompose. Changing crops can also reduce emissions, as can using different agricultural systems, such as organic methods.

Recent work in the US on the energy used to produce different foods has found that meat-eaters are responsible for much greater emissions than vegetarians. An American meat-eater's annual carbon emissions were estimated to be 1485 kg CO_2e more than those of someone on a vegetarian diet – about the same difference between driving a hybrid Toyota Prius and a gas-guzzling 4WD. Eating meat was shown to be responsible for 6 percent of total US greenhouse-gas emissions. Since most US beef is grain-fed, a move to grass-fed meat production would help.

As we saw earlier, stopping all deforestation was equivalent to one stabilisation wedge, and would save 25 Gt C over 50 years. Large areas of the Amazon, African and southeast Asian tropical forests continue to be clear-felled for timber and agriculture, and there is a need for incentives to reduce and hopefully stop this destruction. At the same time, encouraging reafforestation in many parts of the world will help to remove carbon from the atmosphere. The fate of the world's forests will be a key element in post-Kyoto negotiations, and the developed world could find itself paying developing countries to leave their forests standing.

There have also been ideas for controlling global warming that sound like science fiction but still have to be taken seriously. This is the strange world of geoengineering – designing engineering solutions to our climate problem. The basic idea is simple enough. If we can't control our carbon emissions (or, perhaps, don't want to) then why don't we cool the world down by reducing the amount of heat coming in from the sun? Spaceflight engineers suggest it might be possible to construct a gigantic parasol or array of mirrors between the Earth and the sun, to shade the planet and prevent too much warming (see Fig 23, p. 104). If the Arctic ice melts and the much darker ocean underneath starts to absorb too much heat, why not float billions of little white balls on the sea to reflect sunlight back out to space? We know from the impact of volcanoes on climate that sulphate aerosols injected high enough in the atmosphere can cool the planet, so why not use aircraft to deliberately add sulphates to the air? At least one Nobel Prize winner thinks this would be worthwhile. There have been suggestions that it might be possible to extract CO_2 from the air and sequester it underground, and several research teams are already working on ways to capture CO_2 by passing air over chemicals which absorb some of the CO_2 it contains. The CO_2 would then be pumped away and buried, and the chemicals recycled. At the moment these systems are speculative and may be very expensive to operate, but they do have the great advantage of removing carbon directly from the atmosphere. In the very long term, that might allow us to reduce the total amount of greenhouse-gases in the atmosphere, and start a process of global cooling.

The basic problem with most geoengineering solutions is that they involve tinkering with a system we don't understand well. The results of, say, injecting sulphate aerosols over the Pacific will be difficult to predict, vary between regions, and might not all be positive. Rapid cooling over the Pacific could cause damaging knock-on effects all round the world, just as the ENSO has global weather impacts. The prospect of an increase in acid rain as the sulphur eventually comes back to earth isn't attractive, either. At the moment, all geoengineering proposals look fanciful, especially when stacked up against the relatively straightforward technologies that are available to reduce emissions now. The IPCC regards them as "largely speculative and unproven". But if climate change were to speed up, turning science fiction to fact might suddenly seem an attractive option.

10

A low-carbon New Zealand

New Zealand can't afford to ignore the challenge of reducing its greenhouse-gas emissions. Despite the fact that nothing we can do will make a significant difference to climate change here (or anywhere else in the world) we have to play the same game as everyone else. It's very much in our interest that the rest of the world gets on with sorting out the problem, because that is our only hope of minimising the impacts we'll face here. At the same time, the rest of the world is unlikely to look kindly on countries that don't do their bit, however small it might be. Europe, led by France, has already suggested a carbon tariff on goods from any country that doesn't take part in global emissions reductions. Technically, this is to prevent what's called "leakage" – emissions that result from European activity but take place outside the EU's carbon rules. European goods already have a carbon price built in, and so will be more expensive than imports that don't play the carbon game. To prevent that, a leakage charge would attempt to level the playing field and protect European producers from unfair competition. The idea sends a rather direct political signal to countries tempted to cheat on emissions controls in order to get a trading advantage.

New Zealand has long recognised that, as a very small player on the international political scene, the only way it can achieve its national goals is through multilateral negotiation. By being an active partner in the negotiations on post-Kyoto emissions-control schemes, it can influence the eventual result and ensure that its concerns are listened to, if not always acted on.

As we saw in Chapter 4, our geographical position means we will feel the impacts of global warming more slowly than most of the rest of the world. We also have some natural advantages when it comes to reducing our emissions, and one or two disadvantages. Our emissions profile (the mixture of greenhouse-gases we emit) is unusual in global terms because half of our emissions come from agriculture – far higher than the world average of 14 percent. Most of our agricultural gas emissions consist of methane, generated by sheep and cows, and nitrous oxide from fertiliser and dung. The energy sector accounts for the vast majority of the balance of emissions, over 40 percent, divided approximately equally between transport and electricity generation. The balance (a little over 5 percent) is produced by industrial activity (see Fig 24, p. 104).

Under the Kyoto Protocol, each country is entitled to offset its emissions against its carbon sinks – the things the country has that remove carbon from the atmosphere. In New Zealand, forestry is the biggest carbon sink. Until recently, it was thought that this sink would be large enough to achieve our emissions target under the protocol. Trees would suck more carbon out of the air than we emitted from all sources. We would not only be carbon neutral but would have a positive balance of carbon credits which we would sell at a profit. This led successive governments to take a rather relaxed view of the need for action.

Unfortunately, this turned out to be a mistake. New forest planting had been greatly overestimated and the amount being felled was underestimated. To make matters worse, between 1990 and 2004 emissions from power generation increased by over 70 percent, and transport emissions by over 60 percent. This turned New Zealand from a net seller of carbon credits into a buyer. In early 2007 it was estimated that buying enough credits to meet our target would cost the taxpayer more than NZ$650 million. This means, in effect, that because the government hasn't imposed any carbon cost on New Zealand emitters to cover the liability, the taxpayer is subsidising those emissions. Ultimately, the only way to ensure that the market gets the right signals is by exposing emitters to the full cost of their emissions – either through a carbon charge (such as a tax) or through an emissions-trading system. It's also important that the price set on carbon be the international price so that our businesses are operating with the same cost structure as the rest of the world. The Treasury estimated that the world price in June 2006 was NZ$15.92 per tonne of CO_2e, but this could rise substantially during the Kyoto commitment period. If it does, the cost to the taxpayer will rise. Businesses need to know where they stand regarding this major cost, and have been lobbying the government for clear answers. To make long-term investment decisions, they need to know what carbon-pricing systems they're going to have factor in and be confident that the system will be consistently applied over the long term.

Despite the higher emissions and smaller forestry carbon sink, New Zealand remains well placed to achieve Prime Minister Helen Clark's stated goal of carbon neutrality, meaning that the country should have no net carbon emissions. In other words, either our emissions have to be low enough to be covered by our sinks, or the sinks have to be large enough to cover the emissions. Finding a balance between the two is going to be tricky, and designing policies that will get us there a matter for a great deal of highly political juggling. Meanwhile, the rest of the world is moving on. Norway has announced that it intends to be carbon-neutral by 2050. The New Zealand government has announced no targets for emissions or timetable to reach neutrality. At the time of writing, it was undertaking a huge exercise in consultation, putting out policy ideas for discussion by and feedback from, interested parties across many different sectors. The government has indicated that it favours an emissions trading scheme that covers all greenhouse-gases and the whole economy, but trying to work out the likely detail of government policy is very difficult, so here I will confine myself to looking at the broad issues.

Much depends on how strict a view you take of neutrality. Should the country be carbon-neutral in Kyoto terms (i.e. using the strict definitions in the protocol) or can it achieve neutrality by using carbon sinks that are not covered by the agreement? For example, the Department of Conservation estimates that a total spend of NZ$200 million on killing possums and other pests in native forests would allow the recovering forests to remove enough carbon – between 20 and 40 million tonnes by 2012 – to offset all our expected Kyoto shortfall. As our 70 million possums are estimated to eat seven million tones of vegetation a year, it's easy to see how a reduction in their numbers would quickly boost forest regrowth. To me this seems a great idea: combining action on climate change with improving our highly valuable, visible and tourism-friendly conservation estate. And if it can be made to count in the post-Kyoto rules, then so much the better. We could develop a carbon account to go alongside our national economic accounts, recording our carbon emissions/sink balance under both international rules and local standards. If our local offsets are shown to be high quality, they may well earn us goodwill in the same way that our famous "clean green" image has done.

New Zealand can realistically aspire to carbon neutrality because it has a mild climate, a small population, and an energy infrastructure that already uses a high proportion of renewable sources. In 2005 hydro and geothermal power generation produced about 70 percent of the nation's electricity (equivalent to a quarter of the country's total energy use). That has dropped from 81 percent of electricity generation in 1990. Electricity demand has been growing and the shortfall has been met by burning coal and Maui gas. Cheap energy over many decades has reduced the incentive to improve efficiency, but efficiency measures and a shift to renewable generation could make New Zealand's electricity system zero-carbon as early as 2020.

Wind power is already playing a significant role in generating New Zealand's electricity. Not everyone likes the idea of serried ranks of giant windmills spinning on hills around the country, but they are cost-effective and considerable expansion is planned. Over 2007–9 the Ministry of Economic Development expects more than 800 MW of wind generation to be installed around the country. Wind-power schemes qualify for carbon credits under Kyoto rules, and this extra income makes them attractive to electricity companies. The main problem with wind power is that it works only when the wind blows. This limits the amount of wind generation you can have in an electricity network to 20–35 percent, but that still leaves plenty of room for further expansion in New Zealand. There are also restrictions on where wind farms can be sited. They obviously have to be in windy places, and preferably close to the national power grid. They tend to be sited on ridges or ranges, and as a result have a significant visual impact. The negative effects of wind farms in iconic landscapes such as Central Otago, and the public protest that results, suggest electricity companies will have to be very careful with site selection if expansion is to be rapid.

Geothermal power is another low-carbon renewable resource, and there is plenty of scope for an increase in generation. In 2005, the existing

Fig. 25 Meridian Energy's White Hills wind farm in Southland under construction. It will provide enough energy for 30,000 homes. *Meridian Energy*

geothermal stations generated 6.5 percent of New Zealand's electricity. The Ministry of Economic Development estimates this could be quadrupled. Mighty River Power is building a new 70 MW geothermal plant at Kawerau near Rotorua, and Contact Energy has announced it will spend NZ$1 billion on increased generation in the Taupo region. Planning issues will need to be carefully handled, as many prime geothermal sites are in areas of importance to tangata whenua.

The opportunities to expand our hydro capacity are more limited. The large South Island schemes in the Waitaki and Clutha catchments suffer from relatively small lake storage, making the country vulnerable to dry-year power shortages. The environmental impacts on rivers and river flows make new schemes highly controversial, and effectively killed Meridian's Project Aqua, a power and irrigation scheme proposed for the lower Waitaki River. Many major rivers are protected from any further water use by water conservation orders which have been hard won by environmental activism, and it would take a brave government to override these. Future hydro projects are likely to be relatively small run-of-river schemes which would not involve building large storage dams, or improvements to existing schemes such as the recent new Manapouri tailrace. In an ideal system, our hydro lakes would be the "storage batteries" of the national power system, storing water whenever the demand for power could be met from other sources, and generating power when other sources couldn't meet the demand. However, the sale of the power-generation system in the 1990s made that sort of operational strategy unworkable. To make good profits, each of the five generating companies has to sell as much electricity as possible. Leaving water in lakes makes sense from a strategic point of view, but not from a profit perspective.

The potential for large-scale solar-power generation in New Zealand is limited by installation costs, and in some areas by the climate, but that doesn't mean photovoltaic panels won't become more common. In niche uses, such as providing power at sites at some distance from the power grid, they are already economic. Passive-solar heating of household water and buildings is already cost-effective, and being encouraged by the government because of the very significant reduction in electricity demand such systems deliver. A typical household can save up to 40 percent on electricity use with a solar hot-water system. Solar furnaces, using mirrors to concentrate the sun's rays on to a steam generator, are under development, but are not likely to be suited to our climate.

Large-scale use of tidal and wave power looks more promising in the short term. New Zealand has a very long coastline, much of it exposed to big waves, and Cook Strait has very powerful tidal flows. One South Island company has proposed the idea of tidal turbines, like undersea windmills, moored to the floor of the Cook Strait. This sort of technology is still in its infancy but attracts enormous interest, around the world. Scotland, for example, expects to have 1200 MW of tide-and-wave power generation in place by 2020. Another source of renewable power is to burn biomass, as long as it comes from sustainably managed sources.

None of this means that power generation from burning coal and gas is going to stop in the short term, but if the government and consumers send the right signals to the power generators, fewer new thermal power stations will be built. New coal-fired power stations will have to wait for technically proven and affordable systems to capture and store carbon. Nuclear power is also very unlikely to be an option for New Zealand, at least in the short term. Nuclear power stations are large, very expensive, and difficult to integrate into a small electricity system like New Zealand's. They are not good at being switched on and off rapidly, and so are much less flexible than hydro or thermal generation. Smaller reactor designs may be better suited for use here, but for the foreseeable future there is neither the political will nor the necessary infrastructure for New Zealand to go nuclear.

Another idea is for individual homes or buildings to generate as much as possible of their own power from renewable resources such as the sun or wind. At times when they're generating more than they use, the electricity is sold to the grid. When the sun's not shining or the wind's not blowing, they get electricity back from the grid. The technology for these sorts of installations is being developed and used around the world, and the government here is expected to announce technical standards to allow domestic and small-scale systems to connect to the grid in the near future. Another type of small-scale generation called combined heat and power (CHP) improves fuel efficiency, by using waste heat from the generator to heat water or living space. Christchurch company WhisperGen makes a domestic CHP unit which is being trialled in the UK. Its primary use is as a conventional central-heating boiler, burning gas to heat water and run a central heating system, but it also generates electricity which can be used in the house or sold to the grid, thus making far more efficient use of the fuel.

Over the last 20 years New Zealand's per-capita energy use has been increasing, while in most of the developed world the per-capita energy use has been decreasing. Efficiency improvements are not glamorous. It's hard to get excited about changes to the building code, for instance, but that sort of action can make a huge difference to the energy budget. The government announced changes to the building code in May 2007, claiming that new homes would need to use 30 percent less power to achieve a comfortable indoor temperature, saving up to NZ$1800 on power bills per year. The code includes the use of double glazing in most climates, and better insulation. A recent campaign to install "eco bulbs", compact fluorescent light bulbs that are much more efficient than conventional incandescent bulbs, reduced energy demand at a cost of less than 1.5 cents per kWh, compared with building new generating capacity at 6–8 cents per kWh. Providing incentives for consumers to save power in this way not only reducing power bills, but reduces the need to build more power stations – an important reason why Australia and New Zealand are considering phasing out incandescent bulbs. The government has announced new rules on the use of energy-efficient lighting in new and refitted commercial buildings, which it expects will save owners as much as NZ$8 million a year. In coming years, lighting systems based on light-emitting diodes (LEDs) will become cost-effective, offering further efficiency improvements and a wider range of lighting options.

Encouraging the use of passive-solar heating for both domestic and commercial hot-water systems and heating buildings is also an obvious step. The government has announced it will spend NZ$15.5 million to promote solar hot-water systems, and has a target of 20,000 installations by 2010. It should not be difficult to insist that all new buildings should come with solar hot-water systems, but for the time-being the government is content with moves to reduce red tape and make installations cheaper. Financial help for retrofitting solar heating and improved insulation in older homes would reduce electricity demand, and improve health for many.

Efficiency improvements will play a key role in reducing New Zealand's reliance on liquid fossil fuels for transport. We use about 6.3 billion litres of transport fuel a year: about 3.4 billion litres of petrol and 2.9 billion litres of diesel. Increasing the fuel efficiency of motor vehicles will help to decrease the amount of high-carbon fuel used, and by adding biofuel (ethanol or biodiesel) to the blend the carbon intensity of transport emissions can be decreased. The government has already announced it will require total fuel sales to include 3.5 percent of biofuel by 2011. This might not seem like a huge target, but it is limited by the fact that many of our cars are relatively old and by the government's aim of creating a domestic biofuel industry rather than importing more fuel. To address this, it has indicated that it intends to set more stringent fuel-use and economy standards for all vehicles. Replacing gas-guzzling cars in the ministerial fleet with more fuel-efficient and smaller models will also send the right kind of signal.

The Royal Society, in its 2006 report *2020: Energy Opportunities*, projected that New Zealand could be much more radical and aim to obtain all its transport fuel from locally-produced biofuels. This would be a major undertaking, planting and cropping biofuel feedstocks from between two and three million hectares of land (an area of more than 140 × 140 kilometres) that's currently only marginally productive. Wood, preferably from regularly coppiced hardwood trees, could produce ethanol for biofuel and a natural by-product, lignin, to replace fossil-fuel-based substances in plastics, resins, paints and adhesives. An aggressive biofuels programme could bring a significant double benefit, even if complete self-sufficiency is a long way off.

Beyond the use of biofuels, the government is considering encouraging greater use of hybrid and plug-in hybrid vehicles. At present these are most economic around town, where driving short distances on battery power greatly increases fuel-efficiency. However, as we saw in Chapter 9, there are some exciting improvements in battery technology and electric-car design on the way, and if these vehicles are charged by renewable electricity they can be effectively carbon-neutral.

Other transport-related initiatives which can have a significant impact on our national carbon account include encouraging freight off the road and back on to rail. Rail is much less carbon-intensive for bulk transport over long distances. Commuter rail transport is also important in encouraging urban commuters off the road, but is receiving much less funding than road building to relieve traffic congestion. Finding a sensible balance between road-building and public transport is certain to be one of the most debated issues in coming years.

If the political will is there, New Zealand could have a good shot at developing zero-carbon electricity and transport systems over the next 20–30 years. In reality, it seems unlikely that either sector will have to go that far, provided that forestry continues to act as a net sink and the agricultural sector also takes steps to limit its emissions.

Pastoral farming produces half of our total greenhouse-gas emissions, largely in the form of methane and nitrous oxide, two highly active greenhouse-gases. Burping farm animals produce about a third of our total greenhouse-gas output, and agricultural emissions have been growing at one percent a year since 1990. Over the Kyoto commitment period, emissions from this sector are expected to be 38.5 million tonnes above 1990 levels. Although there have been some efficiency improvements (in 1990 a kilo of milk solids produced emissions of about 8.5 kg CO_2e, but in 2004 only 7.5 kg), growth in the dairy industry saw methane emissions rise from 238,000 tonnes in 1990 to 394,000 tonnes in 2002, and it looks as though there is no simple way to achieve reductions.

Methane is a digestion by-product from the stomachs of ruminants. Researchers have looked at changes in diet – altering the ratio of grass species and legumes, for instance – and at changing the mix of stomach bacteria as a means of reducing the gas output, but to date neither approach has produced any dramatic improvement. Interest in this sort of greenhouse-gas mitigation is not high on the list of international research priorities because, globally, agricultural emissions are a small fraction of total emissions. However, work is continuing. Other methane emissions, for example, from working the soil, can be reduced by adopting low-tillage systems.

Nitrous oxide emissions can be reduced by careful monitoring of nutrient levels at the paddock level, and applying only as much fertiliser as can be used by the plants as they grow. Research into nitrification inhibitors (substances that can inhibit the breakdown of nitrogen fertilizers) suggests they can reduce waste by up to 70 percent. This will reduce nutrient run-off into waterways, nitrate penetration into groundwater and greenhouse-gas emissions, and save farmers money. The government has suggested that it might introduce incentives for the use of nitrification inhibitors, and consider the introduction of a tax on nitrogen fertilisers – swiftly dubbed a "fert tax" by farming organisations. It estimates that imposing a 10 percent tax would reduce emissions by 10 percent. The fert tax could fund the inhibitor incentives, and both working together could have a very significant impact on nitrous oxide emissions, though what the cost to farmers might be is not clear. Using inhibitors saves them money because it increases the efficiency of fertiliser application, and applying less fertiliser might offset the increased cost, but the impact will vary from farm to farm.

The government has suggested that methane and general farm emissions might be covered by a tradable-permit regime – a mini-version of the Kyoto cap-and-trade system. Farmers would be free to buy and sell emissions permits and choose the least-cost means of achieving emissions targets. The effect would be to expose farming to the international carbon price. Unfortunately, there is no agreed international standard for measuring farm emissions, and a trading scheme to cover New Zealand's 40,000 farming businesses would be administratively time-consuming. That also makes some of the other policy suggestions more difficult to implement, for example a system of on- or off-farm carbon offsets, or setting national emissions standards that could be applied under the Resource Management Act.

How these ideas will eventually yield a coherent agricultural greenhouse-gas reduction strategy is far from clear. The farming lobby is very influential, as shown by the highly effective campaign against the fart tax, and there is certainly a strand of influential farming opinion that regards emissions-reduction schemes as government interference. One dairy farming leader told a meeting in early 2007 that farmers "should not be held accountable for natural, biological systems". Of course it's natural for a cow to burp, but a dairy farm is in no sense a "natural biological system". On the other hand, many farmers understand that they cannot expect their emissions to escape carbon-pricing, either nationally or internationally. If farming were exempted from the need to reduce emissions, the rest of the country would have to bear the cost of complying with Kyoto, and this would amount to a significant subsidy. At the same time, exports of agricultural products would be at risk of exclusion from prime markets through the farmers being seen as not seeking emissions reductions. This, as we shall see when examining the concept of "food miles", could have a huge impact on our agricultural exports.

The forestry sector has a big role to play in New Zealand's response to climate change. A hectare of mature radiata pine holds about 800 tonnes of carbon, worth about NZ$13,000 in carbon credits at 2006 prices. As long as that forest stands, the money is a credit in the national carbon account. If the forest is cut down, it is assumed that the carbon is returned to the atmosphere and the country has to find NZ$13,000 worth of carbon credits from somewhere else. To qualify for carbon credits, the trees have to be in what's called a "Kyoto forest" – defined as a forest planted on land that was unforested on 1 January 1990, or "Kyoto compliant land". This system of credits was set up under the Kyoto Protocol to encourage countries to use trees as offsets, helping them to meet emissions targets which are tied to that 1990 baseline.

The government needs to find a way to encourage the planting of new Kyoto-compliant forests, making the national carbon sink bigger, and to discourage the felling of trees – especially trees in older, non-Kyoto forests. This has to be balanced against the forestry sector's right to harvest and sell its trees. Not surprisingly, the policy options are being hotly debated, and central to the debate is the ownership of the carbon credits that accrue to the forests. At NZ$13,000 per hectare they represent a considerable increase in income for forest owners, given that a hectare of pruned radiata pine might be worth NZ$25–50,000 after felling costs. From 2007 onwards, people planting new Kyoto-compliant forests will almost certainly be able to get at least part of the financial benefit from the carbon credits their trees earn, but it remains to be seen how that will be allocated. Meanwhile, some forest owners want credits backdated to 1990, which would earn them an enormous windfall profit and be a significant cost to the taxpayer, who would have to fund the resulting shortfall of carbon credits in the national account. To keep the forest industry supplied with timber, the government has proposed a mechanism it calls the Permanent Forest Sink Initiative. Provided that trees are harvested at a rate that allows the forest canopy to remain unbroken – in other words, that the forest is a permanent feature of the landscape – the forest owner will receive carbon credits.

Cutting down forests attracts a penalty under the Kyoto rules, so an important part of government policy is concerned with addressing how that might be administered. This could involve charging flat fees or a tradable deforestation permit regime, but owners will ultimately have to accept some costs associated with removing forests. This deforestation penalty is also highly controversial, as is any potential charge for land-use changes. At the time of writing, forests were being cleared to make way for dairy farms in parts of the North Island and in Canterbury, replacing carbon sinks with potent greenhouse-gas emitters. Landcorp, a state-owned enterprise and New Zealand's biggest agribusiness, is planning to convert 25,000 hectares of forest in the central North Island to dairy and pastoral farming. This is not good news for the national carbon balance, and government policy will have to find a way to discourage that sort of land-use change.

Encouraging beneficial land-use changes could mean more than just encouraging the planting of new forests. One way to cope with climate change is to adapt the crops being grown to the new climate, but that can also be a way to cut overall emissions. Changing from pastoral farming to alternative crops can significantly reduce emissions of methane and nitrous oxide, and at the same time add new carbon sinks. If those crops are also more profitable than dryland pastoral agriculture, everyone benefits. Profitable tree crops, or crops such as truffles and mushrooms that grow in association with the roots of living trees, could be a way to both mitigate and adapt. Farms on marginally productive land could grow feedstock for biofuels. These sorts of land-use changes are good for New Zealand's emissions budget and the economy.

One important opportunity is "carbon farming" to offset other emissions. Carbon offsets are a small-scale version of the Kyoto trading mechanism. They enable individuals or organisations to compensate for their emissions by buying credits from someone who is growing plants to capture carbon or otherwise reducing emissions in a measurable way. By purchasing enough offsets, you can become carbon neutral. This is a useful way to cope with the emissions you can't avoid producing, but

Fig. 26 Converting land from forestry to pasture for dairying in the central North Island replaces carbon sinks with emitters of methane and nitrous oxide. *Landcorp*

they can never be a total solution because there aren't enough offsets available to cover all emissions. Being a low carbon emitter or carbon-neutral can be a very important part of a business response to the challenges of marketing in an emissions-sensitive world. If consumers become concerned about the carbon footprint of the products they buy, then companies that are able to demonstrate their commitment to emissions reductions will have an advantage. This has already worked to the advantage of the New Zealand Wine Company, owners of the Grove Mill brand. Working with Landcare Research's CarboNZero team, they have produced New Zealand's (and perhaps the world's) first wine to be carbon neutral at the point of delivery in their key markets. In the highly competitive UK wine market, this has helped the brand increase sales, with one supermarket, Sainsbury's, doubling their order.

The CarbonZero team is also working with Toyota NZ and South Pacific Pictures, producers of Shortland Street, to help them to manage their carbon emissions. There are three steps: measurement, management and mitigation. Measurement means working out your emissions profile, such as how much electricity and petrol you use. The CarboNZero website has calculators to help businesses, households, schools and tourist operations work out their carbon profiles, and Landcare Research can provide further advice. Management is about taking steps to reduce those emissions, and mitigation means buying offsets to compensate for emissions that can't be eliminated. The CarboNZero scheme therefore ensures that real emissions management is undertaken before offsets are purchased, which is essential if the offset purchases are to be more than "greenwash" – the buying of green credentials without any underlying commitment to the goal of reducing emissions. When it comes to marketing, credibility is everything, and the CarboNZero scheme is designed to deliver that.

Landcare Research also operates its own certified offsets scheme – the Emissions-Biodiversity Exchange (EBEX21). Landowners agree to allow marginally productive land to regenerate as native bush, and are paid for the carbon the new growth takes out of the atmosphere. Sites are usually large (over 100 hectares) and are carefully selected so that it's native bush that grows. The landowners have to fence off the site to keep grazing animals out, remove weeds and control pests. Landcare Research staff measure the site every 10 years and certify the amount of carbon being sequestered. This generates a flow of carbon credits that the landowner can sell in the New Zealand market (often to CarboNZero participants) and provides income from marginal land. Emissions units are measured in tonnes of CO_2e, and in early 2007 were selling for NZ$15 each – a very cost-effective means of approaching carbon neutrality.

Other offsets include carbon credits accrued by renewable energy generation. A new wind farm earns carbon credits because the electricity it produces would otherwise be generated from fossil fuels. These credits are allocated by the government under the Projects to Reduce Emissions scheme, and can be used in Kyoto trading. They provide cashflow for the owners, and have played a large part in encouraging the building of new wind farms around the country. In 2006 Meridian Energy sold between 400,000 and 650,000 emissions reduction units

(ERUs) to the Swiss Climate Cent Foundation from its new 58 MW White Hill wind farm near Mossburn in Southland. The ERUs will be sold at a rate of at least 80,000 per year over 2008–12, the Kyoto commitment period. Capturing methane emissions from landfills and putting them to use can also earn ERUs, and both the Christchurch City Council and the Palmerston North City Council have sold ERUs overseas in the last few years. These sorts of carbon trades are going to become an important part of New Zealand commerce.

Moving towards a low-carbon New Zealand means not only addressing all the major sources of emissions but also making adjustments in our commercial and personal lives. The business community in New Zealand has been divided over the climate-change issue. Some business leaders have been openly sceptical about the reality or magnitude of the problem, while others have moved quickly to adjust to the new global realities. A few years ago, when the government dropped its plans for a carbon charge, the business community was as happy as farmers had been when the fart tax was dropped. Opinion has shifted significantly since then and there is a growing awareness that dealing with climate change is as much about opportunity as it is about cost. Announcing the New Zealand Stock Exchange's move into carbon trading, CEO Mark Weldon said, "Fear tactics and dire predictions of cost impositions are doing nothing to prepare New Zealand businesses for the reality of the future. It's time we stopped talking about threats and started seeing the very real opportunities that carbon trading presents to our businesses, regardless of size, and our national economy". Businesses can see that action is necessary (if only because their export markets are demanding it) and are therefore looking to government to provide clear guidance about the future, and to guarantee a consistent long-term approach. In the short term, schemes to encourage businesses to look at their emissions profiles (the "measurement" step in the CarboNZero programme) make sense as money-saving exercises. Increasing energy and fuel efficiency will add money to any company's bottom line, cutting overheads and reducing emissions at the same time.

At the household level, similar steps can also save money. Changing from incandescent to compact fluorescent light bulbs will be cost-effective. Changing to an electricity supply based on renewable generation will completely eliminate your contribution to carbon emissions caused by power generation, as Meridian Energy has been keen to point out in recent advertising. When looking to change cars, buy a vehicle with low fuel consumption and make sure it can use biofuels when they're introduced. Try to reduce the amount you drive, and try to make each trip as efficient as possible by combining visits or appointments. If buying or building a new home, install solar hot-water systems and insist on good insulation. Buy carbon offsets for any air travel that you do, but take care to buy good-quality offsets. Offsets are not widely marketed here, and low-quality offsets overseas have attracted a lot of criticism because of poor design and lax auditing, but provided that individuals take the same steps as businesses to measure, manage and mitigate, offsets can be worthwhile.

Some of these are small steps, others large, but when you put them all together and get the whole population thinking in those terms, the impact will be significant. The IPCC's Working Group 3 report emphasizes that "changes in lifestyle and behaviour patterns can contribute to climate change mitigation". The power of the consumer to drive business and political change can't be ignored.

11

The big picture

Moving New Zealand towards carbon neutrality is going to be a huge political, technical and economic challenge, but no matter how successful we are, we will be vulnerable to what the rest of the world does. If the international community fails to restrain carbon emissions, the resulting climate change will affect us. At the same time, if our trading partners take actions that make it difficult for New Zealand businesses to sell their products, the domestic economy will suffer. We are vulnerable to both the fact of climate change and the actions taken to combat it.

One of New Zealand's main problems is its location. Being down south in the Roaring Forties may be an advantage from a purely climatic perspective, but it is a liability when doing business with the rest of the world. Since the arrival of European settlers, the country's economy has always been affected by the distance to major markets. If farmers had been able to sell their produce only to the local population, the economy would have been strangled at birth. At first sailing ships took wool to Britain, then refrigerated ships enabled farmers to participate in the international meat market. In the modern global economy, exporters and importers take it for granted that goods and people can be moved quickly all around the world. Only when it is necessary to ship something as cheaply as possible does time cease to matter.

Cheap air and sea transport make globalisation work, but both methods are now coming under close scrutiny for their carbon emissions, which are not covered by the Kyoto Protocol. At the same time, consumers in Europe and the USA have begun to look at the relationship between where their food is grown and the environmental impact of transporting it. The idea of food miles has developed as a quick and easy way to describe that impact. Growing snow peas in Kenya and flying them to Britain in winter has a bigger environmental impact (and produces greater carbon emissions) than growing them in Kent. At the same time, eating locally-produced and in-season produce has become a food fashion, as the huge growth in farmers' markets demonstrates. I like to drink the local Waipara wine and buy food from our farmers' market. Eating fresh, locally-grown and seasonal products is a way to combine the enjoyment of good food with support for local farmers and growers. The concept of food miles is a simple rule of thumb to apply when shopping, and that's one reason why it has become so popular in Europe. It's bad news for New Zealand food exporters, however, because we're so many miles away from our markets. If all that is considered is the distance food travels to get to market, we lose.

The idea of food miles is a blunt instrument – easy to apply, but it doesn't tell the whole story. When a Pacific Rose apple gleams on a supermarket shelf in the UK, the true measure of the global environmental impact of getting it to Hendon or Harrogate is a lot more complex than just the distance it has travelled. The total emissions produced by growing apples in New Zealand and shipping them round the world – the carbon footprint – can be less than for apples grown in Britain or Europe, according to recent research by the Agribusiness and Economics Research Unit at Lincoln University. If you compare the full energy use and carbon emissions of a New Zealand agricultural system with a similar European system, NZ producers can come out on top. Dairy production in New Zealand is twice as efficient as in Europe, and sheepmeat production four times more efficient. Some crops, for example onions, are more efficiently grown in the UK than in New Zealand but still have a larger carbon footprint when the cost of storing them during the off-season is taken into account. The long distance to market is more than offset by our other advantages, such as the mild climate, grass-fed dairy and meat production, and the high percentage of renewable electricity generation.

British supermarkets also recognise that the issue of food miles is a challenge to their business. They rely on being able to display a wide range of fresh produce all year round, and that inevitably means importing produce from far-flung places. Tesco, for instance, has announced that it will label all its products with their carbon footprint, to help customers make environmentally sensible choices. The Carbon Trust, a British organisation set up to help businesses cut emissions, has announced a new carbon-labelling scheme. This measures a product's full emissions from manufacture to use and disposal, to arrive at a figure for the carbon footprint. It made its first appearance on packets of

Walkers Cheese & Onion Flavour Crisps, which have a carbon footprint of 75 g per packet. Tesco is the biggest supermarket business in Britain, and other chains have committed to follow its lead. This is good news for New Zealand growers, but the message won't be lost on Tesco's other international suppliers. If persuading consumers to buy your products means having a small carbon footprint, then producers all round the world will respond to that market signal. We may have an advantage at the moment, but we can't expect it to last forever.

The sensible strategy is to tackle the food-miles issue head-on. Exporters need to emphasise that it's the carbon footprint of a product that counts, and demonstrate that they're working hard at making it smaller. Emissions-reduction schemes such as CarboNZero® certification can make New Zealand products carbon-neutral on arrival at the overseas market.

Encouraging low-carbon shipping would also be sensible. A considerable amount of work has been done on applying modern technology to provide wind power for ships – not a complete return to the days of clippers and schooners, but cargo vessels with computer-controlled aerofoils that are more efficient than traditional canvas, or large kites flying high above the sea where winds tend to be stronger and more reliable. These systems can significantly cut fuel costs and therefore carbon emissions. For a nation with a fine tradition of sailing excellence, investing in wind-augmented cargo ships would make sense in a carbon-constrained world. I'm looking forward to seeing the first New Zealand-built wine clipper arriving in the Port of London – clean, green, carbon-neutral and a fantastic bit of national PR.

If we're vulnerable to food miles, we're also vulnerable to campaigns to reduce what British environmental writer George Monbiot has dubbed "love miles". This is the distance people travel to visit family and friends, and for tourism, one of New Zealand's biggest businesses. In 2006, 2.4 million tourists arrived here, staying a total of 49 million person-nights and bringing a huge amount of overseas earnings into the country. In 2004 overseas visitors spent NZ$8.1 billion, or 18.7 percent of our foreign earnings. Cheap air travel has revolutionised New Zealand's relationship

Fig 27 Trials of the German-designed Skysail system – basically a large kite flown high above a ship, shown here being launched from a test vessel – have shown that it can significantly reduce fuel consumption. *Skysail*

with the rest of the world. But that air travel comes with a carbon cost that is now beginning to be recognised and campaigned against. The British Department for Transport reckons that long-haul flights emit 110 g of CO_2 per passenger-kilometre. On an Auckland to Los Angeles flight that's 1.15 tonnes of CO_2 per passenger; to London, it's more than two tonnes. Unfortunately, jet aircraft also emit lots of water at the top of the troposphere. This forms contrails and cirrus clouds which are very effective at stopping heat from escaping from the surface of the

Earth, but also reflect sunlight before it can reach the ground. When all US air travel was shut down after the 9/11 terrorist attacks, the average daily temperature range increased by 1°C because more heat was able to escape at night, and more sunlight reached the ground during the day. The IPCC reckons that the global-warming effect of aircraft emissions is four times greater than the effect of an equivalent amount emitted at the earth's surface. So while aviation carbon emissions may be only 2.5 percent of the global total, their effect is more like 10 percent.

Aviation is therefore becoming a prime target for emissions reductions. Environmentalists like Monbiot argue that long-haul flying is unethical in a carbon-constrained world. Even the British Conservative party – not normally noted for proposing higher taxes – has suggested it might introduce a "green air miles allowance". Travellers would pay a lower tax rate on one short-haul flight a year, but a higher amount on more frequent or long-haul flights. Tesco plans to label all air-freighted products and limit them to less than one percent of the products in store.

There can be little doubt that eventually aviation will be subject to some sort of emissions control through fuel or ticket taxes or emissions-trading mechanisms. This will make flying more expensive, and while it might not mean a reduction in visitors to New Zealand in the short term, it might spell an end to the steady growth we've seen in recent years. The mix of visitors could tip towards the wealthy end of the spectrum and away from the younger and more environmentally aware "backpacker" market.

If New Zealand can establish a global reputation as a leader in the move to a low-carbon lifestyle and towards carbon neutrality, that will play well with the more eco-conscious prospective visitor, but it will not help us to get around the fact that there are no short-term solutions to the "love miles" problem. The answer, as with food miles, is to confront the issue head-on. For example, in January 2007 Tourism Tasmania developed a promotion with a North American travel group and promoted it as Green Class Travel. For the period of the promotion, all US and Canadian travellers to Tasmania had the carbon emissions from their flight offset up to a value of A$100 through the purchase of credits in offset schemes in North America. This is very clever marketing. It recognises the problem, deals with it, and enhances Tasmania's reputation in an important sector of its target market.

New Zealand's tourism and aviation industry needs to take a similar approach. We can't move the country nearer to the US or Europe, but we can take an aggressive position on travel-related emissions. Air New Zealand could offset its long-haul emissions by investing in local offset schemes. If they were, for instance, to fund the Department of Conservation's possum-control programme, native forest regeneration would provide the carbon offset, while the airline and tourism industry would have an extremely strong selling point. The government could require all international flight arrivals to purchase offsets in New Zealand. This sort of strategic repositioning is the only way we can minimise our vulnerability to the campaigns against long-haul flights that are already underway in Europe. It might be argued that this will simply increase costs, reduce passenger numbers and deter foreign airlines from flying here, but the greater danger is that New Zealand could develop the reputation of being an "environmentally bad" place to go, and this would be very difficult and expensive to change. Better to be proactive and face the problem before it arises. Addressing the tyranny of distance is going to involve further improving New Zealand's international image. We are already seen to be "clean and green", and being carbon neutral is only a logical extension of this.

Some international travel is always going to be necessary, but the rapid rise of the internet in the last decade has changed the way that many businesses relate to their customers. Video and audio conferencing has become much less costly, and in some cases free as computing power and network bandwidth have increased. If physical travel is going to be more expensive and bad for our image, then we need to ensure that our electronic connections are world-class.

New Zealand's other main area of vulnerability has to do with the state of our knowledge about the climate system. At the moment, the evidence suggests that climate change will be moderate and gradual: that we can adapt to the changes that are already inevitable and – if we can cut emissions enough – prevent the worst and most damaging temperature increases. Unfortunately, there is a chance that things won't go so smoothly. As we saw in Chapter 2, there are positive feedback mechanisms that could kick in and boost temperatures far faster than we expect. In particular, carbon-cycle feedbacks could boost atmospheric carbon even if we manage to substantially reduce our emissions. Similarly, an Arctic Ocean that's ice-free in summer could have unpredictable effects on the northern hemisphere climate. These are possibilities for which we have no probabilities: we simply don't know the full extent of the risk. This is the downside of our current knowledge of global warming, and it's a big one.

Once again, we will be at the mercy of how the rest of the world chooses to react. If climate change were to speed up, or a series of high-profile disasters such as Hurricane Katrina caused great economic dislocation, the world might decide that rapid action on emissions was essential. Indeed, there have already been calls for a "wartime response": to redirect national economies to make emissions reduction the primary purpose of all activity. This might seem an unlikely prospect at the moment, and some will argue that it's being alarmist to even raise the issue, but even if it's only faintly possible we ought to consider what strategy New Zealand might adopt. It's one worst-case scenario among many. The first and most important point is that the policies we design to start the process of reducing emissions should be adjustable in response to any greater urgency. If we need to tighten the screw, then we need to be able to do so quickly and with as little economic dislocation as possible. We also need to consider how our economy might respond if international air travel were to be rationed or severely restricted, or if our major overseas markets were to restructure their economies to reduce emissions. If the pace of sea-level rise increases, our Pacific neighbours will naturally look to us for assistance. How will we respond to an influx of climate refugees? It seems unlikely that boatloads of refugees from Indonesia and Asia would try to get as far as New Zealand, but how would we react if they did? Refugees of an altogether different sort – wealthy people looking to escape the worst of the northern hemisphere warming, or expat Kiwis returning home – could bring money but drive up house and land prices. If we are seen to be a good place to escape the worst of climate change, lifeboat New Zealand could quickly become overcrowded. Managing immigration will become even more of a political hot potato if there are thousands of people knocking at the door.

Sustainability has become something of a policy buzzword, but thinking about worst-case scenarios suggests that we need to put national sustainability into a world context. New Zealand depends on the healthy functioning of global trade. Many of the things we rely on, such as vehicles, computers and pharmaceuticals are made overseas. There is no realistic prospect that factories will spring up here to meet demand if some of the goods we need to import become hard to get. We are therefore vulnerable to anything that restricts our ability to trade, whether here or overseas. We can do our best to develop greater self-reliance, but full self-sufficiency will never be possible in a complex modern society – at least not with the standard of living we now take for granted. It might be sensible for governments to look for ways to encourage the retention of basic manufacturing skills in New Zealand, and to encourage industries that are not wholly dependent on large-scale exporting. It will never be a complete answer, but with luck the worst case may never happen.

12

The way forward

Writing this book has been both reassuring and frightening. Reassuring, because if all goes well New Zealand is as well placed as any country in the world to ride out gradual global warming. We can cope, adapt, perhaps even thrive as the climate changes, provided that those changes are not sudden, and don't bring with them dramatic economic dislocation in the rest of the world. The frightening stuff is what happens if the climate system experiences rapid change. What we know about the history of Earth's climate tells us that global and regional climates can change rapidly – over a few decades, perhaps a few years, possibly even in as little as one season. But the climate record doesn't help us much in working out the risks of future sudden change because we don't have a similar period in the past to guide us. To paraphrase former US Secretary of Defense Donald Rumsfeld, we have left the realm of the "known unknowns" (the things we are aware that we don't know) and are heading out into the "unknown unknowns" (the surprises). Our output of greenhouse-gases and our impact on the surface of the planet have pushed the climate well out of the normal range over the last four million years, and we don't yet know what that might mean. We're boldly going where no man has gone before.

Our best guess, based on work by thousands of scientists and summarised by the IPCC, is that the global climate will change gradually. If the governments of the world can work together to produce the necessary reductions in greenhouse-gas emissions, and as a result cap temperature increases at less than 2°C, then the world as a whole can probably cope. Of course, 2°C is a global figure. Some places will get much warmer, particularly at high latitudes, and that has implications for sea ice in the Arctic, and ice loss from Greenland and Antarctica. Ultimately, the sea level will rise to accommodate that ice melt, but if we're lucky that will be a slow enough process to adapt to. If we end up heading for more than 2°C (and the pessimist in me believes we probably are), it could be rough ride. I think it would be better, on balance, to be more aggressive in our actions now and do our best to avoid that risk.

At the same time, I am a great believer in humanity's ability to gather knowledge and develop technologies that will help us cope

Trace Hodgson

with a warming world. In my lifetime slide rules have made way for supercomputers, man has landed on the moon and sent robots to explore the solar system, and we have decoded the human genome. I am therefore optimistic about our ability to create a low-carbon world that offers a high standard of living for all its inhabitants. I'm a technological optimist. But there is a race on between our increasing scientific prowess and the damage climate change is doing. If we're lucky, we'll get the job done before the damage gets too great. My optimism is tempered by climate pessimism. If we're unlucky, global warming could be bad enough to make developing advanced technologies the least of our concerns. We have to both adapt to and mitigate global warming. We also have to innovate and implement mitigation technologies.

Whatever form the post-Kyoto emissions-reduction deals take, I would like to see a parallel global research-and-development (R&D) push to develop relevant low-carbon technologies. It would take only a small fraction of current global military R&D expenditure to provide a huge boost to work in key areas such as solar energy, batteries, biofuels and bioengineering for carbon capture, which if introduced quickly and shared between the developed and developing world could improve our chances of stabilising and then lowering atmospheric greenhouse-gas levels. Helping developing countries to meet emissions targets also makes sense and would be a useful way to encourage them into the post-Kyoto fold.

There is also need for a significant increase in research into global warming and climate change. The IPCC has demonstrated that the problem is here, and that it poses a great threat. We now need to know more about the climate system, and in particular the oceans and ice sheets, to see if we can detect the signs of rapid, damaging change on the way. We also need good medium-term forecasts of climate, and greatly improved regional information to inform our adaptation efforts. An early-warning system for damaging climate change would be a very worthwhile goal.

These are all sensible ways to approach the problem of global warming and the climate change it causes, but beyond the weather statistics and flood probabilities there is an even larger issue that our civilisation will have to confront. There are about 6.5 billion people on the planet today. By the middle of the century there will probably be 9 billion, and a third of them will live in China and India. Those two nations are developing at breakneck speed. If all their citizens are to achieve the standard of living we enjoy in our little corner of the South Pacific, or the resource use that North Americans take for granted, can the planet cope?

Since 1987, an organisation called the Global Footprint Network has been calculating what it calls World Overshoot Day. Overshoot Day is the day in each year when the world starts to live beyond its ecological means – when human consumption of resources exceeds the biosphere's annual production. When we overshoot, we start withdrawing capital from our planetary account. The first Overshoot Day was 19 December 1987. In 2006, Overshoot Day was 9 October. We are using about 30 percent more of our natural resources than the biosphere can regenerate. We are not using the planet sustainably.

It's possible to dispute these figures, to argue about the calculations and what they represent, but the underlying message is clear. With 6.5 billion people, we are already running down our natural capital. As the rest of the world develops, this problem can only get worse. Those of us who are lucky enough to live in the developed world have no right to insist that the unfortunate in Africa or Asia should be condemned to poverty while we live high on the hog. The people of China and India have every right to aspire to our standard of living. There has to be a levelling-up of living standards, or there will be a levelling-down forced on us by climate change, resource depletion, the unsustainable mining of the biosphere, and the social and economic disruption those things will bring. From this perspective, climate change is just one symptom of the bigger underlying problem: the impact of the huge numbers of human beings on the planet.

In New Zealand and in the developed world, we take it for granted that economic growth is a good and necessary thing. Our economies have been built on the exploitation of natural resources, and the assumption that if one source runs out another can be found to replace it. As the concern about peak oil demonstrates, that assumption is now being challenged. Facing up to ecological overshoot means that we have to look again at how we can grow without depleting the planet's biosphere. Economic growth is going to have to be sustainable growth. That will mean rethinking the way we run the global economy, and will be an intellectual and political challenge for our leaders. Being a technological optimist I believe we can do it, but I am certain it will not be easy.

In New Zealand, we are better off than the planet at large. Our small population and relatively large land area suggest that we can live within our ecological means. If we are careful, we can meet the sustainable-growth challenge within our own borders. But, as with climate change, we are vulnerable to what happens everywhere else. For the time being, it is enough that our national debate should focus on our response to global warming because that will point us in the right direction, but as the passage of time sharpens our focus on the problem of global overshoot we will need to be prepared for even more fundamental change.

As a small country, we are well equipped to adapt to rapid change. We can change our political and economic course very quickly, as the reforms of the 1980s and 1990s demonstrated. We can do the same again in addressing climate change, but to do that we must first arrive at a national consensus on the need for action. We have to recognise the problem and look for ways to adapt to change and reduce emissions. Our politicians must face up to the problem and find the will to act. Political will can exist only with voter support, so popular opinion needs to stiffen our parliamentarians' backbones.

Learning to live with and trying to limit climate change is going to drive national and international politics over the next century. Carbon policy therefore has to be at the heart of government policy, not an addition or afterthought. Nor can fine words about carbon neutrality replace action to get us there. If we are to thrive in the changing world, we need to be seen to be playing our part, even taking a lead. Either that, or we become an irrelevance on the world scene, perhaps even an outcast.

As I said in my introduction, science provides us with the information we need to act but doesn't tell us what to do. The politics is in the response we choose to make. It's solutions we need, not specious debate about whether the problem even exists. In this book I have tried to convey the scale of the problem we face and sketch out some of the potential solutions, but I have not attempted to say exactly what New Zealand must do. That's a complex subject, and a debate we need to have. This book is my contribution.

If we can lead the world towards carbon neutrality, it will be good for business, good for our people and good for the country. When it comes to dealing with climate change, we are all people of this land. It is our privilege to live here, our duty to hand the land on to future generations in good condition.

APPENDIX

The sceptical view

This appendix examines the views of the vocal minority who maintain that climate change isn't real or worth doing anything about. It was originally intended to be a chapter in the middle of the book, to follow my discussion of climate science as we understand it today. But as the book neared publication, it became obvious that the sceptics were becoming increasingly irrelevant to the climate debate in New Zealand. How can you make a sensible contribution to a debate on emissions-control policy or energy strategy if you start by denying that there's a problem worth addressing? Once the balance of debate had moved on from the existence of the problem to looking for solutions, the sceptics became irrelevant. That shift seemed to happen with surprising speed over the end of 2006 and into 2007, and I have therefore moved "their" chapter to the back of the book. An interesting afterthought, a curiosity, a matter of passing interest, but of no relevance to the future of our country.

When the issue of global warming first became a matter of concern in the late 1980s, many people were prepared to deny both the scientific basis of the problem and the need for any action. Credible climate scientists could argue that CO_2's warming effects would probably be minor, or that there were negative feedbacks in the climate system that would limit its effect. This was not what most scientists thought, but it was at least plausible.

Those concerned by the prospect of global warming and who believed that there might be a need to reduce carbon emissions could find encouragement in the success of the Montreal Protocol, an international treaty negotiated in the 1980s to reduce the emissions of chlorofluorocarbons (CFCs), the chemicals that cause the ozone hole. Science had found a problem, warned of its effects, and governments had (eventually) acted to restrict CFC use. Companies producing CFCs had been upset at first, warning of dire impacts on their profits and the economy at large and lobbying to resist controls, but in the end accepted the need for action, losing little money in the process.

If global warming was a problem, then commercial interests in the fossil-fuel business such as giant oil companies, coal miners and car producers could see there was a risk that the same thing might happen.

Any action to restrict CO_2 emissions would dramatically change their businesses. In 1989, a group of US and multinational companies (including Exxon Mobil, Shell, BP, Texaco and most of the oil business, Daimler Chrysler, Ford and General Motors) formed a lobby group called the Global Climate Coalition (GCC), and over the next decade spent millions of dollars trying to persuade governments to do nothing. At the same time, global environmental groups such as Greenpeace and the World Wide Fund for Nature began to actively campaign in the opposite direction. Between the two lobbies, climate scientists got on with their work, but science and politics were becoming inextricably intertwined.

Through the 1990s, as the science progressed and IPCC reports found increasing evidence of warming with a greenhouse signature, it became more difficult for sceptical scientists to argue convincingly that a CO_2 build-up was having no effect. Although there was (and still is) plenty of debate about how the planet's climate system works, and how it might change in the future, there is now little argument about the basics. Nevertheless, the GCC and other corporates, working through a network of sponsored scientists and so-called "think tanks" (which are really pressure groups rather than *bona fide* researchers), tried to promote the idea that there was still an active debate about the reality of global warming. In a memo circulated in 1998 as part of a response to the Kyoto Protocol of the preceding year, a GCC spin-off group calling itself the Global Climate Science Team outlined a PR and communications strategy to promote uncertainty about the science underlying global warming: "Victory will be achieved when those promoting the Kyoto Protocol on the basis of extant science appear to be out of touch with reality." This tactic of promoting a spurious or trivial debate in order to undercut the need for action was pioneered by the tobacco industry in the 1970s and, in the US at least, it worked. Although the Clinton administration actively participated in the Kyoto negotiations (Clinton's vice-president Al Gore flew to Kyoto at the last minute to broker a deal when the process looked like failing), the US refused to ratify the accord, and President George W. Bush's administration has explicitly rejected joining it in the future. The GCC, however, folded in 2002, having lost most of its major sponsors, most of whom were worried about the PR damage they might sustain by being seen to be arguing against the strong evidence presented by the IPCC's 2001 Third Report.

The Kyoto Protocol was (and remains) anathema to many on the right of the political spectrum, especially that strand of American (and world) politics that advocates free-market libertarian policies as the solution to all society's ills. This also feeds into a well-established anti-environmentalist strand in that school of political thought. In some quarters, believing that global warming presents a serious problem is tantamount to being a tree-hugging greenie, intent on imposing socialism on the world. In Australia, the Howard government refused to ratify Kyoto, and until quite recently was openly sceptical of the need for any real action. In New Zealand, the National Party under Don Brash was equivocal about the need for action, and advocated withdrawing from Kyoto.

Over the last couple of years, however, there has been a dramatic shift in attitude to climate policy by centre-right parties around the world. In Britain, which has always been more receptive to the reality of the problem, the Conservative Party under David Cameron has put climate change at the centre of its policy platform. In a similar if more modest move, the National Party under John Key has committed itself to fulfilling the 2012 target set out in Kyoto, and set a 2050 target of reducing net carbon emissions to 50 percent of the 1990 level. In the USA the Democrats' sweep of the 2006 Senate and Congress elections suggests that global warming and climate change are going to be back on the policy agenda. The political reality has changed.

Sceptical arguments

There is still no shortage of people who will argue that climate change is not worth worrying about. Their motives are sometimes political, based on the sort of anti-Kyoto views discussed above, and sometimes rooted in a belief that the science is all wrong. There are a range of "climate sceptic" views, from outright denial of the existence of global warming (which may include insisting that the world's been "cooling since 1998"), to admitting that there's some "modest" warming (but it's not caused by humans, therefore we don't need to do anything) or even that, if warming is real and we are the cause, it won't do much damage. One characteristic of sceptics' arguments is that they start from their chosen position, and then select only facts or theories that support it. They ignore or downplay evidence that doesn't suit their position, while promoting even the most tenuous arguments that suit their views. As the saying goes, "If you hear hoofbeats in the night, think first of horses, not zebras". Climate sceptics are fond of zebras, preferring to ignore (to mix my animal metaphors) the "elephant in the room" – the effect of greenhouse-gases on the planet's climate.

The use of the word "sceptic" for those who argue against the reality of climate change and the need for action does a disservice to the honourable tradition of scientific scepticism. Scientists have to be prepared to challenge new ideas, to demand proof and to test theories. Then, if the theory stands up, they have to accept and adopt this new way of thinking in their work. Climate sceptics are good at the first part, but struggle with the second.

New Zealand has its own collection of sceptical voices, who have banded themselves together as the New Zealand Climate Science Coalition (NZCSC). They say they are "concerned at the misleading information being disseminated about climate change and so-called anthropogenic (man-made) global warming. The Coalition is committed to ensuring that New Zealanders receive balanced scientific opinions that reflect the truth about climate change and the exaggerated claims that have been made about anthropogenic global warming". The NZCSC included recently deceased TV weatherman Augie Auer and Queensland geologist Professor Bob Carter, and has also welcomed British TV personality and vociferous climate-change denier David Bellamy into its ranks. The NZCSC writes letters to the New Zealand press, distributes press releases on climate issues, and plans to "audit" the IPCC's Fourth Assessment Report. In its launch "position statement", the NZCSC says it is time to "challenge the IPCC's monopoly on the 'official view' of climate change information and advice to governments". Given that very few of the NZCSC's members are climate scientists and none are publishing relevant material in the peer-reviewed scientific literature, it is difficult to see why any responsible government would consider their advice relevant.

In their announcement that David Bellamy was joining the coalition (issued in October 2006), they described what they called their "seven pillars of climate wisdom". These "pillars" amount to a sceptic's catechism, so it's worth taking a look at each one in turn.

1. Over the last few thousand years, the climate in many parts of the world has been warmer and cooler than it is now. Civilisations and cultures flourished in the warmer periods.
This is both true and irrelevant. There is a lot of natural climate variability in the system, but we have pushed one of those variables – the concentration of CO_2 in the atmosphere – well beyond the levels seen at any time since the recurring sequence of ice ages began.

2. A major driver of climate change is variability in solar effects, such as sunspot cycles, the sun's magnetic field and solar particles. These may account in great part for climate change during the last century. Evidence to date suggests warming involving increased CO_2 exerts only a minor influence.
There is no doubt about the sun's role in driving our climate system, but it acts through direct radiation, not through sunspot cycles, magnetic fields or by modulating galactic cosmic rays. There has been no significant increase in solar radiation since the 1950s, during which time the Earth's temperature has increased by 0.6°C. A variation in the sun's output may have accounted for some of the warming experienced over the early part of the 20th century, but the recent warming cannot be explained without invoking the effects of increased amounts of greenhouse-gases. The claim that sunspot cycles, the solar magnetic field and solar particles are "a major driver" of the climate is not supported in the scientific literature.

But it tells us something about the sceptics: since they won't accept the role of increasing greenhouse-gases, they have to find another explanation. The sun's an obvious possibility, but nobody has yet found evidence for the kind of links they claim. On the other hand, CO_2's role as a greenhouse-gas is well understood. Atmospheric CO_2 has increased by more than a third since the mid-1800s, and that this will cause the planet to warm is basic physics.

We also have good evidence for what's called a "greenhouse signature" in the temperature record. After sunset, the day's heat is radiated out into space and the Earth's surface cools down. If greenhouse-gases trap this heat, you would expect nights to become warmer than they used to be. This can be seen in the temperature record. It should also mean that winter months are less cold than they used to be and, again, this is seen to be true.

3. Since 1998, global temperature has not increased. Projection of solar cycles suggests that cooling could set in and continue to about 2030.
This is an example of "cherry-picking": choosing data that suit you in order to make a point and ignoring data that don't. But it's still not a good example, because it's transparently wrong. Why? First, in order to make this claim the sceptics have to choose 1998 as their starting point, and use the Hadley Centre dataset for world temperatures. (If they used the NASA dataset, it would show 2005 was slightly warmer than 1998, but that's inconvenient for their argument so they ignore it.) Second, 1998 was an unusual year: the real reason it was much warmer than 1997 or 1999 was because a large El Niño was present in the Pacific. But if you make a graph from the Hadley Centre data, it is obvious that 2005 was only slightly cooler than 1998, even though it wasn't influenced by a large El Niño (see Fig 29, p. 104).

Third, most of the warmer years are at the top end of the graph. If you plot a trend line through the data points they show an upward trend – not flat or downward. In fact, it makes very little difference to the trend whether you start the graph in 1997 or 1999. The year 1998 was what's called an "outlier", a data point significantly above or below the trend line. The telling point is that what was an unusually warm year in the late 1990s is more or less normal eight years later.

The "cooling to 2030" claim is based on one paper, not published in a scientific journal, which has found no support from anyone except climate sceptics.

4. Most recent climate and weather events are not unusual: they occur regularly. For example, in the 1930s the Arctic region experienced higher temperatures and had less ice than it has now.
Not true. Parts of the Arctic certainly had a warm spell in the 1930s, but it was not as widespread or as warm as today, and in the 1930s there was much more ice covering sea and land. It might have been possible to make this claim in the 1990s, but the Arctic has carried on warming. It's a good example of a "zombie fact" that refuses to lie down and die long after it has been shown to be no longer true.

5. Stories of impending climate disaster are based almost entirely on global climate models. Not one of these models has shown that it can reliably predict future climate.
Data on rapid climate change come not from climate models but from the paleoclimate record, which shows sharp swings from warm to cold and back again when the Earth warmed as ice ages ended. These swings were larger than we've experienced in recent centuries. What the models show is that the world will *continue* warming as we add more greenhouse-gases to the atmosphere. How warm depends on how much gas. In general, climate models don't project the sort of rapid changes that have occurred in the past. And they have made some good predictions. In Chapter 2 we saw how a 1988 vintage climate model gave a very good idea of the warming that would occur over the next 15 years.

6. The Kyoto Protocol, if fully implemented, would make no measurable difference to world temperatures. The trillions of dollars that it will cost would be far better spent on solving known problems such as the provision of clean water, reducing air pollution and fighting malaria and AIDS.
Kyoto was never intended as a complete solution to the climate-change problem. It was just a first step, to put in place mechanisms and markets

that would enable each country to arrive at a least-cost way to reduce its own emissions. The costs of implementing Kyoto are nowhere near as large as the sceptics say (realistic estimates are discussed in Chapter 9). Comparing the costs of dealing with climate change with other problems is confusing two separate issues: switching to a low-carbon economy does not have to mean ignoring AIDS, malaria or pollution.

7. Climate is constantly changing and the future will include coolings, warmings, floods, droughts and storms. The best climate policy is to make sure that we have in place disaster-response plans that can deal with weather extremes and react adaptively to longer-term climate cooling and warming trends.

Climate does change, and will continue to change. The world certainly will experience floods, droughts and storms in the future, but the chances of significant long-term cooling in the near future are remote, barring large changes in solar output or a lot of very dramatic volcanic activity. The "best" climate policy is a matter for debate, but it will certainly have to include being ready to deal with weather extremes and adapting to change. For the time being, it makes more sense to plan for rising sea levels and a warmer climate than to prepare for glaciers to come marching out of the alpine valleys.

The NZCSC's "seven pillars of climate disinformation" are relatively easy to debunk by anyone who has taken the time to become familiar with the science of climate change. Unfortunately, most of the general public and politicians haven't. It's easy to assume that sceptical arguments are a valid counter-balance to an IPCC view, that there is a continuing debate about the reality and severity of the problem. "The science is not settled" is the climate sceptics' clarion call. And if they say it often enough, and confidently enough, they will fool some of the people, some of the time. The principal weakness of the sceptics' position is that they start from the wrong end of the problem. They want to encourage us to believe there is reasonable doubt, like defence lawyers addressing a jury in a courtroom. It doesn't matter that their theories may not hold any water, or that their evidence is illusory. If they can create enough doubt to prevent action, they've won.

NZCSC chairman Rear Admiral Jack Welch recently provided a good example of all these techniques when he launched a popgun broadside against Auckland mayor Dick Hubbard and Waitakere mayor Bob Harvey. In May 2007 the two mayors issued a press release about the need to educate their communities on climate change, describing it as "the monster in the living room". Welch rushed out a rebuttal release to chastise them for their efforts. He wrote: "The really monstrous reality is that leaders such as the two mayors are rushing to get on a global warming bandwagon for which there is no valid verifiable scientific proof. The first thing they should check is New Zealand's official temperature and sea level records, where they will find that the country has been cooling since the El Niño of 1998, and the levels of the Waitemata Harbour have remained about the same for the past 100 years."

Two sentences. Many errors. "No valid verifiable scientific proof" of global warming? Only if you ignore the entire IPCC Fourth Report, which finds the evidence for global warming to be "unequivocal". The Rear Admiral then refers to data from New Zealand, implying that if we aren't getting warmer, then global warming must be in doubt. But New Zealand isn't the whole world, so the situation here says little about the rest of the planet. He then indulges in some cherry picking, by using the "cooling since 1998" trick we looked at above, and then takes one sea-level measurement and implies that it can disprove the national and global average figures. He conveniently ignored the fact that since 1950 New Zealand has warmed by 0.4°C and sea levels have risen by 70 mm, as the IPCC reported.

Unfortunately for sceptics like the Rear Admiral, the world's real climate scientists are not actively debating the basic science underlying our understanding of the problem. There's plenty of interesting and sometimes heated debate about the details, but the basic message has been explicitly endorsed not only by the vast majority of climate scientists working through the IPCC, but also by most of the world's science organisations (including the Royal Society of New Zealand) and leading Earth science academic institutions. The basic science *is* settled. To prove otherwise, the laws of physics would need to be rewritten, and there's no sign of a new Einstein among the current crop of climate-change deniers.

New Zealand's sceptics occasionally feature in our media, but they are not major players on the world scene. In the USA, the disinformation strategies put in place a decade ago are still playing out. According to the Union of Concerned Scientists (UCS) in a report published in early 2007, Exxon Mobil has continued to fund a campaign to mislead the public about the true dangers of climate change. Between 1998 and 2005 the company spent US$16 million through a network of "think tanks", websites and hired "experts" to spread disinformation. It's easy to find the results on the internet, or in the US press.

As the world continues to heat up it will become more difficult for sceptics to claim that the problem doesn't exist – at least if they want anyone to believe them. "Professional" sceptics in the USA are already adopting the view that there is "some modest warming" and that human emissions *might* be playing a role. However, they are still pushing the "science isn't settled" line, and continuing to disparage Kyoto. Meanwhile, the Bush administration is doing its best to slow down international negotiations for a successor to Kyoto. Exxon Mobil must regard its money as well spent.

At some point climate scepticism becomes outright denial, and when denial flies in the face of all evidence, like members of the Flat Earth Society going on a trip to the International Space Station, the sceptics become a laughing stock. Credible scepticism is now restricted to questioning *how much* climate change there's going to be, and how much damage it will cause. That feeds into the policy debate, rather than denying any need for one. But there will always be a few cranks around ready to challenge the causes of warming, even if they may soon have to do it from a rowing boat at a balmy North Pole.

APPENDIX

Notes and resources

An expanded version of this section, with clickable links, can be found at the *Hot Topic* website: www.hot-topic.co.nz

On the *Hot Topic* blog you can discuss any of the material in the book, and keep up with developments in climate science and New Zealand policy news.

Using the internet

There is a vast amount of information available on the internet about global warming and climate change, and a huge range of opinion. It pays to check your sources carefully. For news, I recommend *New Scientist* (www.newscientist.com/home.ns) and *Scientific American* (www.sciam.com). Google News (news.google.co.nz) is also useful, especially if you set up a standard search like "global warming" and register for a personalised Google News page. Google also offers a daily "news alert" service, based on the news search of your choice. For coverage of the latest in climate science, and a continuing effort to correct misrepresentations of that science, RealClimate is without peer: (www.realclimate.org). It's a blog run by working climate scientists, and has very active and interesting discussions in the comments to each post.

Reading scientific papers

Sometimes it's best to go straight to the horse's mouth, but reading scientific papers can be a difficult exercise for someone outside the relevant field. They are not designed to be easy reading, and assume a level of background knowledge of the literature that can be very hard for any non-specialist to achieve without intensive study. Nevertheless, I would recommend that anyone who is half-way scientifically literate give it a try from time to time. Some papers, particularly those that set out to review a field, can provide a very good overview. The IPCC reports (www.ipcc.ch) are excellent examples of straightforward coverage of complex issues. They don't hide the complexities, so they're not bedtime reading, but they do encapsulate the extent of current knowledge.

Chapter 1: Living in a greenhouse

Good explanations of the greenhouse effect can be found on Wikipedia: www.wikipedia.org The relevant articles there are contributed and maintained by climate scientists.

The figure for the global energy imbalance (0.85 ± 0.15 Wm^{-2}) comes from Hansen et (*Science* 308, 3 June 2005), popularly known as "the smoking gun" paper. The energy imbalance is the "smoke". This paper can be downloaded from Hansen's website at Columbia University: www.columbia.edu/~jeh1 and is a fairly easy read, as these things go.

For a much more detailed treatment of the history of climate science, Spencer Weart's *The Discovery of Global Warming* is a superb resource. It is available as a book, website and free downloadable PDFs: www.aip.org/history/climate/index.html The website provides around 250,000 words on the subject. It may fairly be described as comprehensive.

A global climate model that you can run on your home computer, EdGCM, is hosted by Columbia University: edgcm.columbia.edu It will run on most recent home computers, and is not a toy. It's used for teaching and research.

"A different world" is James Hansen's phrase, used to describe the situation where global temperature increases more than 1°C above present levels. He explained why in a talk to the American Geophysical Union, commemorating Keeling, in 2005: www.columbia.edu/~jeh1/keeling_talk_and_slides.pdf

The GISS temperature data can be accessed (and downloaded) at data.giss.nasa.gov/gistemp You can see a wide range of graphs and representations of the data by clicking on the graphs link. The Hadley Centre data is here: www.metoffice.gov.uk/research/hadleycentre/obsdata/globaltemperature.html with more at the University of East Anglia's Climatic Research Unit: www.cru.uea.ac.uk

More background on El Niño and ENSO can be found on Wikipedia: en.wikipedia.org/wiki/El_Nino

NIWA provides information on ENSO's impacts on weather in New Zealand here: www.niwascience.co.nz/ncc/faq/ensonz

NOAA in the USA publishes regular ENSO monitoring information here: www.ncdc.noaa.gov/oa/climate/research/2007/enso-monitoring.html

Chapter 2: The climate system

Spencer Weart's *The Discovery of Global Warming* (see above) includes an excellent essay on the development of GCMs.

An introductory overview of the physics of climate modelling by GISS modeller (and RealClimate blogger) Gavin Schmidt is here: www.physicstoday.org/vol-60/iss-1/72_1.html

Met Service rural forecasts are here: www.metservice.com/default/index.php?alias=ruralmap

Hansen writes about his testimony and the controversy that later surrounded his projections, here: www.giss.nasa.gov/edu/gwdebate

The figure for global carbon emissions in 2005 comes from the Global Carbon Project: www.globalcarbonproject.org and the conversion factors for CO_2 to carbon from the Carbon Dioxide Information Analysis Centre at the Oak Ridge National Laboratory in the USA: cdiac.ornl.gov/pns/convert.html

You can read more about the coupled carbon cycle work being done by the Hadley Centre at www.metoffice.gov.uk/research/hadleycentre/models/carbon_cycle/index.html

Figures for carbon release are from the Hadley Centre and the Global Carbon Project.

The figures for committed warming and time to equilibrium are taken from the Hansen "smoking gun" paper, linked above.

Chapter 3: The state of the science

The IPCC website is at: www.ipcc.ch The complete AR4 reports are available for download (free), and if you want to dig deep into the science of climate change there is no better place to start.

James Hansen's Keeling Lecture at the American Geophysical Union's 2005 conference provides an excellent discussion of how climate sensitivity can be worked out from ice-age to interglacial climate change. The arithmetic is shown on slide 12 (linked above).

There's a good map of New Zealand's coastline during the last ice age at Te Ara, The Encyclopedia of New Zealand: www.teara.govt.nz/ EarthSeaAndSky/OceanStudyAndConservation/SeaFloorGeology/3/ ENZ-Resources/Standard/1/en

The State of the Cryosphere site (nsidc.org/sotc) mentioned earlier includes a good discussion of glacier mass balance, as well as an overview of the world's ice and snow.

RealClimate posted a good overview of the then current work on Greenland ice in March 2006: www.realclimate.org/index.php?p=267

The Wikipedia page for the Jakobshavn Isbrae (en.wikipedia.org/wiki/Jakobshavn_Isbræ) includes satellite pictures showing the retreat of the calving front.

The full IPCC Special Report on Emissions Scenarios is available on the web here: www.grida.no/climate/ipcc/emission/index.htm

All the numbers used in this chapter are from the Summary for Policymakers of the Working Group One AR4 Report, which can be downloaded from the IPCC website. It is a very valuable overview of the state of the science, and is cross-referenced with the much more detailed full report, released in May 2007.

Chapter 4: The outlook for New Zealand

Much of this chapter draws on work by NIWA prepared for the government's climate change office as guidance for local authorities. The full range of material is available here: www.climatechange.govt.nz/resources/local-govt/guidance.html

If you'd like to have a preview of what rising sea levels might mean, this website flood.firetree.net uses Google Maps and allows you to choose a sea-level rise (from 1–14 m) to see what happens to any part of the world.

Chapter 5: Impacts: the good and the bad

All dairy industry statistics are from Fonterra: www.fonterra.com/content/dairyingnz/thedairyindustry/default.jsp

Estimates for pasture growth increases are taken from *Climate Change: Likely Impacts on New Zealand Agriculture* (Kenny, 2001) available from the government climate-change website. This report gives a good overview of impacts on the major agricultural sectors.

An excellent overview of the likely impacts on the Gisborne district is *An overview of climate change and possible consequences for the Gisborne district* by Louise Savage (2006), prepared for the Gisborne Civil Defence Emergency Management Group. It can be downloaded from the Gisborne District Council here: www.gdc.govt.nz/NR/rdonlyres/ ECDB7664-9AA8-4EA0-BB8B-2AF837BA7C98/0/ ClimateChangeReport2.pdf It is a model that other regions would do well to emulate.

Chapter 6: Sinking or burning: our Pacific neighbours

Figures for the Pacific Islands are taken from *Climate Trends & Projections for Small Islands: Tropical South-West Pacific* by Penehuro Lefale in *Confronting Climate Change*, edited by Chapman, Boston and Schwass (Victoria University Press, 2006)

Other material is drawn from the IPPC AR4 WG2 summary and report.

Chapter 7: Warming in the wider world

Most of the information in this chapter is from the AR4 WG2 summary.

Another useful resource is *Impacts of Global Climate Change at Different Annual Mean Global Temperature Increases* by Rachel Warren, Chapter 11, in *Avoiding Dangerous Climate Change* (Cambridge University Press, 2006). It provides an excellent overview of the impacts literature, scaled to different global temperature increases. The full text of *Avoiding Dangerous Climate Change* is available as a free download here: www.defra.gov.uk/environment/climatechange/research/dangerous-cc/index.htm

The Stern Review also includes a good section on expected climate change effects, as you might imagine in a report devoted to estimating their economic impacts.

Chapter 8: Facing up to the inevitable

The list of farm adaptation options comes from work conducted by Dr Gavin Kenny for the Ministry for the Environment over 2003/4. You can find the full document here: www.climatechange.govt.nz/resources/

local-govt/adapt-climate-change-eastern-nz-jul05/adapt-climate-change-eastern-nz-jul05.html

For further information on water issues, including irrigation, dairying and contamination, a good plain-language source is the Water Rights Trust website: www.waterrightstrust.org.nz This site deals specifically with Canterbury but the principles it addresses apply throughout the country.

Chapter 9: Cooling the future

The United Nations Framework Convention on Climate Change provides the international support for climate change negotiations. You can find out more here: unfccc.int/2860.php

James Lovelock's arguments can be found in his book *The Revenge of Gaia* (Allen Lane, 2006).

My comments on the probability of hitting the EU's 2°C target are based on *What Does a 2°C Target Mean for Greenhouse Gas Concentrations? A Brief Analysis Based on Multi-Gas Emission Pathways and Several Climate Sensitivity Uncertainty Estimates* by Malte Meinshausen, Chapter 28, in *Avoiding Dangerous Climate Change* (Cambridge University Press, 2006). (Linked above)

The term "procrastination penalty" was first used by a poster at RealClimate, and explored by Bill McKibben in *Warning on Warming*, in *The New York Review of Books* (February 2007): www.nybooks.com/articles/19981

The Stern Review on the Economics of Climate Change is available as a free download here: www.hm-treasury.gov.uk/independent_reviews/stern_review_economics_climate_change/sternreview_index.cfm

The quotations are from the Executive Summary, which gives a very concise and readable overview of the issue.

Socolow and Pacala first described their wedge concept, and gave examples of 17 feasible wedges, in *Science* in 2004. You can download that original article and learn more about wedges at the Princeton website: www.princeton.edu/~cmi/resources/stabwedge.htm

British environmentalist and writer George Monbiot's book *Heat* (Allen Lane, 2006) includes an excellent section on how buildings can be designed to reduce their carbon footprints.

The high fuel efficiency Loremo car project is here: www.loremo.com/
The Tesla Roadster can be found here: www.teslamotors.com/
The figures on carbon emissions associated with a meat-eating diet are from *Diet, Energy and Global Warming* by Gidon Eshel and Pamela Martin, in *Earth Interactions* (Vol 10, 2006). It can be downloaded here: geosci.uchicago.edu/~gidon/papers/nutri/nutri.html

Chapter 10: A low-carbon New Zealand

The full text of the Kyoto Protocol is available here: unfccc.int/resource/docs/convkp/kpeng.html

The welter of discussion documents on climate change-related issues can be found at the government's climate-change website here: www.mfe.govt.nz/publications/climate

Much of the material I draw on for the discussion of NZ's energy options comes from *2020: Energy Opportunities* (a report by the Energy Panel of the Royal Society of New Zealand, August 2006) and is available here: www.rsnz.org/advisory/energy

The Royal Society's suggestions for action are considerably more aggressive than those in the government's own *Draft NZ Energy Strategy 2006*, available here: www.mfe.govt.nz/publications/climate

The government's discussion document on *Sustainable Land Management and Climate Change* is also available from the above site.

The farming leader quoted is Frank Brenmuhl, chairman of Dairy Farmers of New Zealand, as reported by the NZPA on 21 February 2007 in an item headlined *Dairy farmers spit the dummy on greenhouse gas accountability*.

Landcare Research's CarboNZero® certification scheme website: www.carbonzero.co.nz/index.asp

EBEX21 carbon offset information: www.ebex21.co.nz

Chapter 11: The big picture

The Lincoln University study *Food Miles – Comparative Energy/Emissions: Performance of New Zealand's Agriculture Industry*, by Caroline Saunders,

Andrew Barber and Greg Taylor (Research Report No. 285, July 2006), can be downloaded here: www.lincoln.ac.nz/story_images/2328_RR285_s6508.pdf

Tesco's announcement of its green initiative was made in a January 2007 speech by CEO Sir Terry Leahy: www.tesco.com/climatechange/speech.asp

The Carbon Trust (www.carbontrust.co.uk/default.ct) labelling scheme was announced at the end of February 2007: www.carbon-label.co.uk

The Walkers Crisps website has more information: www.walkerscarbonfootprint.co.uk

A *New Scientist* article called the *The New Age of Sail* (www.newscientisttech.com/channel/tech/mg18524881.600-the-new-age-of-sail.html), from 2005, includes details of two approaches to using wind power to improve fuel efficiency in ships. The SkySail kite system is now going into service: www.skysails.info/index.php?L=1

All the tourism statistics come from Statistics NZ : www.stats.govt.nz/economy/industry/tourism.htm

The 2004–5 numbers come from Tourism Satellite Account 2005.

The post-9/11 increase in the US daily temperature range is reported here: www.sciencenews.org/articles/20020511/fob1.asp

Chapter 12: The way forward

Full details of the Global Footprint Network's Overshoot Day and how it is calculated are here: www.footprintnetwork.org/gfn_sub.php?content=overshoot

Appendix: The sceptical view

There is a significant quantity of "sceptical" global warming material on the net, some of it on sites explicitly or covertly funded by fossil-fuel companies and other interested parties, others run by interested individuals.

Full details of membership and actions of the Global Climate Coalition can be found here: www.sourcewatch.org/index.php?title=Global_Climate_Coalition

The Union of Concerned Scientists' report on Exxon Mobil's funding of global warming denial, which includes the full text of the memo from the Global Climate Science Team, can be downloaded from their website: www.ucsusa.org

The New Zealand Climate Science Coalition maintains a website promoting climate change denial, linking to many other such sites: www.climatescience.org.nz

Pillar 2: RealClimate calculates the contributions of the various constituents of the atmosphere to the overall greenhouse effect here: www.realclimate.org/index.php?p=220

Pillar 3: The Hadley Centre global temperature data can be downloaded here: hadobs.metoffice.com/hadcrut3/diagnostics/global/nh%2Bsh/index.html I plotted the graphs by importing the data to a Microsoft Excel spreadsheet and using the "apply trendline" command, selecting a linear trend.

Pillar 4: You can track daily and seasonal changes in sea-ice extent at *The Cryosphere Today*, run by the University of Illinois (arctic.atmos.uiuc.edu/cryosphere) and including excellent animated graphics of sea ice at both poles. For more information (including ice cap, permafrost and glacier data) see the *State of the Cryosphere* site (nsidc.org/sotc) operated by the US National Snow and Ice Data Center. The NSIDC also offers an overlay for Google Earth which tracks northern hemisphere snow cover, permafrost extent, and sea ice.

The press release from the two Auckland mayors is here: www.scoop.co.nz/stories/PO0705/S00103.htm

The NZCSC reply is here: www.scoop.co.nz/stories/SC0705/S00020.htm

An excellent source of information when looking to counter the arguments used by sceptics is a series of articles prepared by Coby Beck under the title *How to Talk to a Climate Sceptic*: illconsidered.blogspot.com/2006/02/how-to-talk-to-global-warming-sceptic.html also available at Gristmill: gristmill.grist.org/skeptics

LIST OF ACRONYMS

CFC	Chlorofluorocarbon
CHP	Combined Heat and Power
EBEX21	Emissions-Biodiversity Exchange
ENSO	El Niño/Southern Oscillation
ERU	Emissions Reduction Units
ETP	Evapotranspiration
GCC	Global Climate Coalition
GCM	Global Climate Model
GISS	Goddard Institute for Space Studies
IPCC	Intergovernmental Panel on Climate Change
NIWA	National Institute of Water and Atmospheric Research Ltd
NZCSC	New Zealand Climate Science Coalition
RCM	Regional Climate Model
SRES	IPCC Special Report on Emissions Scenarios
UNFCCC	United Nations Framework Convention on Climate Change

GLOSSARY

Albedo: The fraction of solar radiation reflected by a surface or object. Snow and deserts have a high albedo, reflecting a lot of the incoming radiation, while oceans and forests have a low albedo. The Earth's albedo averages out at about 30 percent.

Biomass: (1) The total mass of living organisms in a given area or volume. (2) Live or recently harvested material used as a feedstock for energy production.

Carbon footprint: A measure of the greenhouse-gas emissions resulting from an activity, expressed as the weight of CO_2 or carbon.

Carbon neutral: An activity is carbon neutral if it produces no net increase in atmospheric carbon. Any carbon emissions produced are "offset" by activities designed to remove the same amount of carbon from the atmosphere. Zero-carbon activities produce no carbon emissions, and so do not require any offsetting.

Climate commitment: The amount of future climate change resulting from a given level of greenhouse-gases in the atmosphere over the time the planet takes to regain its energy balance (see p. 36).

Climate sensitivity: The amount of climate change that results from a doubling of the amount of greenhouse-gases in the atmosphere (see p. 41).

Climate system: The interrelationship between the atmosphere, hydrosphere (oceans and water), cryosphere (ice and snow), the land surface, and biosphere (all living things).

CO_2 equivalent: A measure of the forcing effect of the greenhouse-gases in the atmosphere, expressed as the amount of CO_2 required to produce the same effect (see p. 48).

CO_2 fertilisation: The increase in plant growth caused by increasing levels of CO_2 in the atmosphere.

Downscaling: The process of taking information from global climate models and relating it to local and regional climates. **Dynamic downscaling** uses a regional climate model fed with the GCM output to produce climate projections. **Statistical downscaling** develops empirical and statistical relationships between global and regional climates (see p. 46).
Emissions trajectory: The change in greenhouse-gas emissions over time as the result of human activity.
Equilibrium response: The amount of climate change that results from a given level of greenhouse-gases when the climate system has achieved energy balance. See also **Climate commitment, Transient response**, and p. 36.
Evapotranspiration: The loss of water to the atmosphere from plants (transpiration) and soil (evaporation), usually expressed as the amount of rain required to replace it.
Forcings: The external drivers of change in the climate system. These include solar variations, changes in the Earth's orbit and changes in greenhouse-gas levels caused by human activities (see p. 30).
Greenhouse-gas: A gas in the atmosphere that absorbs long-wavelength radiation and thus acts to retain heat at the surface of the planet (see p. 16).
Heat engine: A device (in this case the climate system) that converts thermal energy into work as heat flows from a hot region to a cold region.
Holocene: The geological period from about 11,600 years ago until the present.
Insolation: The amount of solar radiation reaching the Earth's surface.
Model: A mathematical description or system of equations that represents a process (see p. 24).
Polar amplification: The prediction by global climate models that warming in the polar regions will be greater and occur more quickly than elsewhere (see p. 43).
Stabilisation wedge: An action to reduce greenhouse-gas emissions that starts slowly but builds over time (see illustration, p. 103). As developed

by Socolow and Pacala, a typical wedge reduces emissions by 25 Gt C over 50 years.

Temperature proxy: A means of measuring temperatures of the past, for example analysing the gas composition of bubbles found in ice cores.

Topoclimate: Detailed mapping of regional and local soil, landforms and climates to provide improved information about crop suitability.

Transient response: The amount of climate change at a given time and given amount of greenhouse-gases, before the climate system has had time to achieve energy balance. See also **Equilibrium response**.

Tropopause: The boundary between the troposphere and the stratosphere (see p. 29).

Troposphere: The lowest layer of the atmosphere, containing 95 percent of the total weight of gases. All weather takes place in this layer (see p. 29).

Detailed explanations of many more climate terms can be found at RealClimate (www.realclimate.org/index.php/archives/category/extras/glossary), and the IPCC Working Group One report includes all the standard definitions (ipcc-wg1.ucar.edu/wg1/Report/AR4WG1_Annexes.pdf).

ACKNOWLEDGEMENTS

In writing *Hot Topic* I have relied on the help and guidance of many people. New Zealand's climate science community has always been ready to provide advice and assistance, often at short notice and at a time when the IPCC report process was demanding much of their attention. David Wratt at NIWA was instrumental in helping me move *Hot Topic* beyond the idea stage and into production, and has helped with contacts, ideas, and comments on much of the text. Jim Renwick, also at NIWA, has helped with information and graphics, and has been kind enough to provide a most generous foreword. Other NIWA scientists who have helped me to get a grasp of the subject are Brett Mullan, Sam Dean, Mike Harvey and Antony Gomez. Gavin Kenny of Earthwise Consulting provided a lot of useful information. Trevor Chinn was kind enough to make sure that my comments on New Zealand's glaciers were accurate, and also dug into his photo archive for the magnificent pair of pictures showing the Tasman glacier in retreat. Professor Blair Fitzharris pointed me at much useful information on the impact of climate change on New Zealand, and Penehoro Lefale did the same for the impacts on Pacific islands. Gavin Schmidt at GISS was kind enough to review my chapter on modelling. If I have got my science straight, it is thanks to these kind people. Any errors are my own.

I'd also like to acknowledge the assistance of Stewart McKenzie and Larry Burrows at Landcare Research, Joseph Arand and Ferne McKenzie at DOC, business commentator Rod Oram and Erick Brenstrum of MetService. They provided me with useful facts, ideas and interesting perspectives on the issue.

I'd like to thank Meridian Energy for the photo of the White Hills wind farm under construction, Landcorp for the dairy conversion picture, Tim Whittaker for the coastal erosion shot, the *Northern Advocate* for the Kerikeri flooding picture, *The Press* for the dairy cows and irrigation shots, the Earth Observatory at NASA for the shot of Australia burning, Gary Braasch for the Tuvalu picture, Tesla Motors for the picture of their cars, Craig Potton for the shot of Alpine vegetation zones on Mt Taranaki, the Department of Conservation for the tuatara picture, the University of Arizona's Steward Observatory for the "solar shield" image, and SkySails for the shot of their trial sail in action. Trace Hodgson provided the telling cartoon in the final chapter.

Special thanks to Roger Smith at Geographix Ltd who provided the base map for the illustration showing the distribution of climate change impacts on New Zealand. I'd like to also thank Robert A. Rohde of the Global Warming Art web site (www.globalwarmingart.com), whose superb graphics provided the base for five key illustrations. NIWA helped out with the graphs showing New Zealand temperatures and Baring Head CO_2 concentrations, and the IPCC gave kind permission to use two graphics from the Summary for Policymakers of the Working Group One report.

Finally I would like to thank the team at HB Media Ltd who put the book together for AUT Media. Mike Bradstock directed the project and edited the book with his customary vim and vigour, Adrian Clapperton art directed and did the cover design, Louise Cuckow designed the body of the book, and publishers Martin Bell and Vincent Heeringa steered it to the bookstalls with enthusiasm. A good book needs a great publishing team. *Hot Topic* has been lucky to have one.

INDEX

A
abrupt climate change, risk of, 89
acidification of oceans, 51
adaptation, 92–4
    time scales for, 93
    in developing nations, 94
adaptive capacity, 94
Africa, 88
agriculture
    impact of climate change, 69–76
    risk of flooding, 69
    impact of drought, 69–70
    impact of GW in developing world, 87–8
    loss of production, 89
    farm level response to GW, 113-4
    emissions, 131-2, 138-9, 145
    emissions trading, 146
aircraft emissions, 157–8
albatross, royal, 67
Alliance of Small Island States, 83
Amazon, 35–6, 88
Andes, glaciers in decline, 88
Antarctic Circumpolar Current, 54
Antarctic Peninsula, 55
Antarctica, 44–5, 63
apricots, 73
arable farming, 73
Arctic, 86
    warming in 1930s, 176
Arrhenius, Svante, 22
Auckland, 75, 76, 112
Auer, Augie, 173
Australia, 13
    effects of warming, 78–81
    banana crop, 80
    sugar cane, 80
    aviation, 129
    avocado growing, 74

B
Baring Head, 24
barley, 73
battery technologies
    flow, 31
    lithium, 131
    ultracapacitors, 131
Bay of Plenty, 57
beach loss, 61
beech forest, 67
Bellamy, David, 173
biofuels, 129, 143
biomass as source of energy, 141
biosecurity, 75, 94
BP, 14
Branson, Sir Richard, 129
Brash, Dr Don, 172
building code, 142
Bush, President George W., 171

C
Callendar, Guy, 23, 25
Cameron, David, 172
Canterbury, 56, 58, 95
cap-and-converge, 124
cap-and-trade, 122–3
carbon account, 138
carbon capture and storage, 127, 142
carbon charge, 137
carbon credits, 147-8, 150
    from forestry, 147–8
carbon cycle, 32–4
carbon cycle modelling, 35
carbon dioxide
    pre-industrial level, 20
    temperature records and, 28, 31–2
    seasonal fluctuation, 31
    emissions from human activity, 33, 116
    atmospheric levels, 41
    fertilisation, 70, 87, 90
carbon dioxide equivalent, 47–8, 120

current level, 120
   possible targets, 120–1
carbon farming, 148–9
carbon footprint, 155
carbon labelling, 155–6
carbon neutrality, 138
carbon offsets, 148, 152
carbon price, 137
carbon tariff, 136
carbon tax, 11, 116, 122
carbon trading, 116
   NZ market, 151
carbon, fossil deposits, 34
CarboNZero, 149-51, 153
Carter, Professor Bob, 173
CCS, *see* carbon capture and storage cement, 34
Central Otago, 73
CFCs, *see* chlorofluorocarbons
cherry-picking, 175
Chile, 54
China, 88
chlorofluorocarbons, 170
Christchurch City Council, 151
Christchurch, 76
citrus growing, 74
Clark, Helen, 13, 138
clathrates, 34, 35
climate, 28–31
climate change, rapid, 176
climate commitment, 36, 58
climate corridor, 65
climate history, of NZ, 54
climate models, 24–5, 30–2, 46
   projections from, 42
   scenarios tested on, 47–8
   regional, 47
climate refugees, 160–1
climate responses, 36
climate sceptics, 170–9
climate sensitivity, 41
CO_2, *see* carbon dioxide
coal-fired power generation, 141–2
coastal erosion, 60
cold stress, effect on death rates, 76
combined heat and power, 142
commuter rail, 144

compact fluorescent lightbulbs, 128, 143
Conservation, Department of, 138
Conservative Party, 158, 172
Contact Energy, 140
Cook Islands, 82
coral, bleaching due to warming, 79
Coromandel, 57
cropping, 73
cyclone of 1936, 60

D
dairying, 70, 95
deforestation, 132
   tradable permits, 148
dengue fever, 75–6, 94
distributed generation, 127, 142
downscaling, 46–7
drought, 73, 94, 107–8
Dunedin, 65
dynamic downscaling, 47, 55

E
Earth Summit, 117
EBEX21, 150
eco bulbs, 143
economic growth, 167
ecosystem services, 68
ecosystems
   response to warming, 64
   response to season timing, 67
   simplification, 68
   services, 68
   impacts as temp increases, 87
   Australian impacts, 88–91
   adaptive capacity of, 94
El Niño, 19, 62, 79, 82, 112, 175
electric cars, 130
   batteries for, 131
emissions
   agricultural, 131–2, 145
   air transport, 157–8
   domestic reductions, 151–2
   from fossil fuel, 125
   land use change, 125
   aviation, 132
   NZ agriculture compared to European, 155

historical, 123
tradable permits, 146
Emissions-Biodiversity Exchange, 150
Emissions Reduction Units, 150
emissions reduction
   cost of, 124–5
   wedges, 126
   in transport, 131
   shipping, 156
   need for flexibility, 160
emissions trading, 122, 137
   see also cap-and-trade
emissions trajectory, 37, 52, 121
energy efficiency
   in housing, 128
   in NZ, 142
energy infrastructure in NZ, 139
ENSO, 82
   see also El Niño/La Niña
equivalent, *see* carbon dioxide equivalent
erosion, 57,107
   coastal, 111–2
ERU, *see* Emissions reduction units
Europe, impacts of global warming, 88
European Union, 120
   target to limit climate change, 120
evapotranspiration, 95
Exxon Mobil, 179

F
farming, see agriculture
fart tax, 146
feedbacks
   positive, 160
   carbon-cycle, 160
fertiliser ("fert") tax, 145
Fiji, 83
fire risk, 69–70
fishery yields, 87
flooding, 95, 107
   risk increases, 69
   Asian deltas, 87
   designing infrastructure
      to cope with, 108
Fonterra, 70, 95
food miles, 147, 154
forcings, 30

forecasting, 29
   medium range, 112

forestry, 147
   as carbon sink, 137
fossil fuel
   extent of deposits, 34
   emissions, 125
   for transport in NZ, 143–4
Fourier, Joseph, 22
Fox glacier, 58
France, nuclear power generation in, 126
Franz Joseph glacier, 58
fuel cell, 130
fuel efficiency, vehicles, 129

G
Gaia hypothesis, 118
gas-fired power, 127
GCM, *see* global climate model
geoengineering, 132–3
geothermal power, 126
   in NZ, 139–40
Gisborne, 56, 75
GISS, *see* Goddard Institute
   for Space Studies
glaciers, 21, 44–5
   in NZ, 58–60
   Andean, 88
   Himalayan, 88
global circulation model, *see* climate models
Global Climate Coalition, 171–2
global climate model, *see* climate models
Global Climate Science Team, 171
Global Footprint Network, 166
Goddard Institute for
   Space Studies (GISS), 18, 39
Gore, Vice President Al, 171
grandfathering, 123
grape growing, 74–5
Great Barrier Reef, 79
greenhouse effect, 16–7
Greenland, 44–6, 89
Greenpeace, 171
groundwater, salt water penetration, 60, 93
Grove Mill wines, 149

global warming
  cause of, 10
  projections for end of century, 49
  pattern of warming, 50
  predicted global effects, 50
  rate of warming, 50
  ecosystem response, 64–69
    see also ecosystems
  changes in seasons, 67
  effect on death rates, 76, 80
  effects on Australia, 78-81
global impacts, 86-90
  impact on water resources, 86–8
  projections of warming, 86
  effects in Europe, 88
  social and political response, 89–90
  as intergenerational issue, 92
  human perception of, 92
  local and regional government response, 93, 107
  healthcare system response, 94
  farm response, 113–4
  need for regional information, 112
  wine industry expansion, 112
  land use change as response, 113
  response levels, 119
  pest control as response to, 138
  impact of aircraft emissions, 158
  wartime response, 160
  research to mitigate, 165
  early warning system, 166
  as symptom of overpopulation, 167

H
Hadley Centre, 18, 39, 175
Hansen, James, 32
Harvey, Bob, 178
Hawke's Bay, 56, 73, 74, 82
heat related deaths
  in NZ, 76
  in Australia, 80
  in North America, 89
heat wave, European of 2003, 86
heather, 64
Himalayan glaciers, 88
  see also glaciers
Holland, 111

Hubbard, Dick, 178
human activity, CO_2 emissions caused by, 33
Hurunui river, 70
hybrid vehicles, 144

hydroelectric generation, 106, 126
  potential in NZ, 140–1
hydrogen, as transport fuel, 130

I
ice age, 20
  warming from, 50
ice core studies, 19–20
India, 88
insolation, 17
interglacial period, last, 51
Intergovernmental Panel on Climate Change (IPCC), 25–6, 40, 86, 166, 173, 175
  Working Group reports, 40
Invercargill, 65, 75
IPCC, see Intergovernmental Panel on Climate Change
irrigation, 57, 72, 96

K
Kakadu National Park, 78
Kawerau, 140
Keeling curve, 24
Keeling, David, 23
Kenya, 154
Kerikeri, 73, 74
Key, John, 172
kikuyu, 72
  see also subtropical grasses
Kiribati, 82
kiwifruit, 73–4
Kyoto emissions targets, 117
Kyoto forest, 147
  see also Forestry
Kyoto Protocol, 11, 117–8, 171, 175
  Clean Development Mechanism, 117–8
  criticisms of, 121
  principles of, 124
  costs of action, 124

L

La Niña, 19, 62
　Manapouri, 141
land use change
　as response to GW, 113
　to reduce emissions, 148
Landcare Research 149, 150
Landcorp, 148
landslides, 58
land-use change, 73
leakage, 136
LED lighting, 143
lignin, 144
Lincoln University, Agribusiness & Economics Research Unit, 155
Little Ice Age, 21
Loremo, 129
love miles, 156
Lovelock, James, 118

M

maize, 73
　as feedstock for biofuels, 129
malaria, 75
Manawatu, 56
Marlborough, 56, 57, 73
Medieval Warm Period, 21
Meridian Energy, 150, 152
methane, 34, 35, 72, 145
MetService, 30
Mighty River Power, 140
migration, 161
　animal, 67
　as response to GW, 83
Milankovitch cycles, 20–1
Ministry of Economic Development, 140
mitigation, 116–133
model, regional climate, 47
modelling, *see* temperature, global circulation, carbon cycle, climate model
mohua, 68
moisture deficit, 57, 96
Monbiot, George, 156
Montreal Protocol, 42, 170
mosquitoes, 75–6
Mossburn, 151
Mt Aspiring National Park, 64
Mt Cook, 58
Muriwai, 75
Murray Darling Basin, 78–9

N

Napier, 112
NASA, 175
National Party, 172
Nauru, 82
Nelson, 74
New South Wales, 80
New Zealand
　emissions profile of, 12
　sea level rise, 60–1
　Kyoto emissions target, 117
　emissions profile, 136–7
　energy infrastructure, 139
　geothermal power, 139–40
　distance to markets, 154
　vulnerability to climate change, 154
　vulnerability to change overseas. 159–60
New Zealand climate
　temperature history, 19
　climate modelling, 47
　projections, 55–61
　drought, 56–8
　rainfall, 56
　rate of warming, 54
　regional changes, 56–8
　variability, 54
　winter warming, 56
New Zealand Climate Science Coalition, 173–79
New Zealand Stock Exchange, 122, 151
New Zealand Wine Company, 149
nitrate pollution, 95, 96
nitrification inhibitors, 145–6
nitrogen fertilisers, 96
nitrous oxide, 72, 145
Niue, 82
NIWA, 31, 109
　uses dynamic downscaling, 47
NZ climate projections, 55–61
Northern Territory, 78
Northland, 57, 73
Norway, 138
nuclear power, 126
　potential in NZ, 142

O

ocean acidification, 51, 116
  effects on coral, 79
ocean, thermal expansion of, 51
Otago, 56, 58
ozone, 42–3

P

Pacala, Stephen, 126
Pacific Islands, 13
  effects of warming, 81
  sea level rise, 82
painted apple moth, 75
Palmerston North City Council, 151
paspalum, 72
  see also subtropical grasses
passive solar heating, 141
PassivHaus, 128
pasture, response to warming, 72
Patagonia, 54
peak oil, 167
Pegasus Bay, 61
permafrost
  melting of, 21, 35
  loss in Himalayas, 88
Permanent Forest Sink Initiative, 147
Peru, 88
pest control as response to GW, 138
pests, spread south, 65, 75
photovoltaic cells, 127
  see also Solar power technologies
photovoltaic panels, 141
plug-in hybrid vehicles, 144
polar amplification, 43, 55
possum control, 138, 159
procrastination penalty, 121
Project Aqua, 140
Projects to Reduce Emissions, 150

Q

Queensland, 78, 80

R

rail transport, 128, 144
rainfall, increase in intensity, 56, 94
Rakaia river, 70
Rangitata river, 70
reafforestation, 132
refugees, climate, 87, 89, 160–1
regional climate modelling, 47
research & development, as response
  to GW, 165
rivers, recreational use, 106
rockfalls, 58
rodents, 68
Ross River virus, 75–6
Royal Society of New Zealand, 144, 179
Rumsfeld, Donald, 164

S

Sainsbury, 149
salt marsh mosquitoes, 75
scenarios, in modelling, 47–8
sceptics, 170–79
sea level change, 44, 45, 51, 89, 95, 164
  NZ impacts, 110
  dealing with, 110–2
shipping, 130
  sail-assisted, 156
ski industry
  in NZ, 58
  in Australia, 80–1
snow line, permanent, 58, 64
snow cover, 58, 64
  decline in snow pack in
    North America, 88
Socolow, Robert, 126
solar hot water systems, 128, 141, 145
solar power technologies, 127
  generation potential in NZ, 141
solar radiation, as driver
  of climate change, 174
South Pacific Pictures, 149
Southeast Asia, 88
Southern Oscillation, 19, 62
  see also El Niño, ENSO
Southland, 56, 80
statistical downscaling, 46, 55
Stern Review, 124
storms, 58
  damage in USA, 89
storm surge, 60, 82
subtropical grasses, 65, 72
sustainability, 161
Sydney, 80

T

Taranaki 56, 80
Tasman Glacier, 58–60, 112
Tasmania, 80
Taupo, 140
temperature data, global, 17–8
temperature modelling, 23, 30
temperature proxy, 18
Tesco, 155–6, 158
Tesla Roadster, 130–1
thermal expansion of oceans, 45
thermal power generation, 106
tidal and ocean current power, 127
    potential in NZ, 141
Toafa, Maatia, 83
Tongariro National Park, 64
topoclimate, 109
tourism, 138
    impacts, 107
    impact of aviation restrictions, 158–9
Toyota NZ, 149
tree crops, 73–4
tropical storm tracks, 57–8
troposphere, 29
trout fishery, 106
tuatara, 66
Tuvalu, 82
Tyndall, John, 22

U

ultracapacitors, 131
Union of Concerned Scientists, 179
United Nations Framework Convention on Climate Change, 11, 117

V

vegetation zones, 64–5
Victoria, 80
vine crops, 73–4
Virgin Airways, 130

W

Waimakariri River, 61, 78
Waipara River, 61
Wairarapa, 57
Waitakere City, 178
Waitaki River, 106, 140
Walkers cheese & onion crisps, 156
water resources
    pressure on, 70
    storage schemes, 70
    decline of, 86–8
    increase in high latitudes, 86
    North & South America, 88
    availability and use, 94–106
    efficiency of use, 105
wave power, potential in NZ, 141
weather, working of, 28–9
wedge concept, 126
weeds, spread south, 65
Welch, Rear Admiral Jack, 178
Weldon, Mark, 151
Wellington, 112
West Antarctica, 89
West Coast, 37, 56
westerly winds, increase in, 56
Western Australia, 79
wetlands
    drainage of, 68
    restoration, 107
Whangarei, 57
wheat, 73
WhisperGen, 142
White Hill wind farm, 151
willow sawfly, 75
wind power, 126, 139
wind, damaging, 57
wine industry
    response to warming, 74–5
    expansion as result of GW, 112
wood, as source of biofuel, 144
World Climate Conference, 25
World Overshoot Day, 166–8
World Wide Fund For Nature, 171

Y

yellowhead, 68

Z

zero-carbon energy system, 139